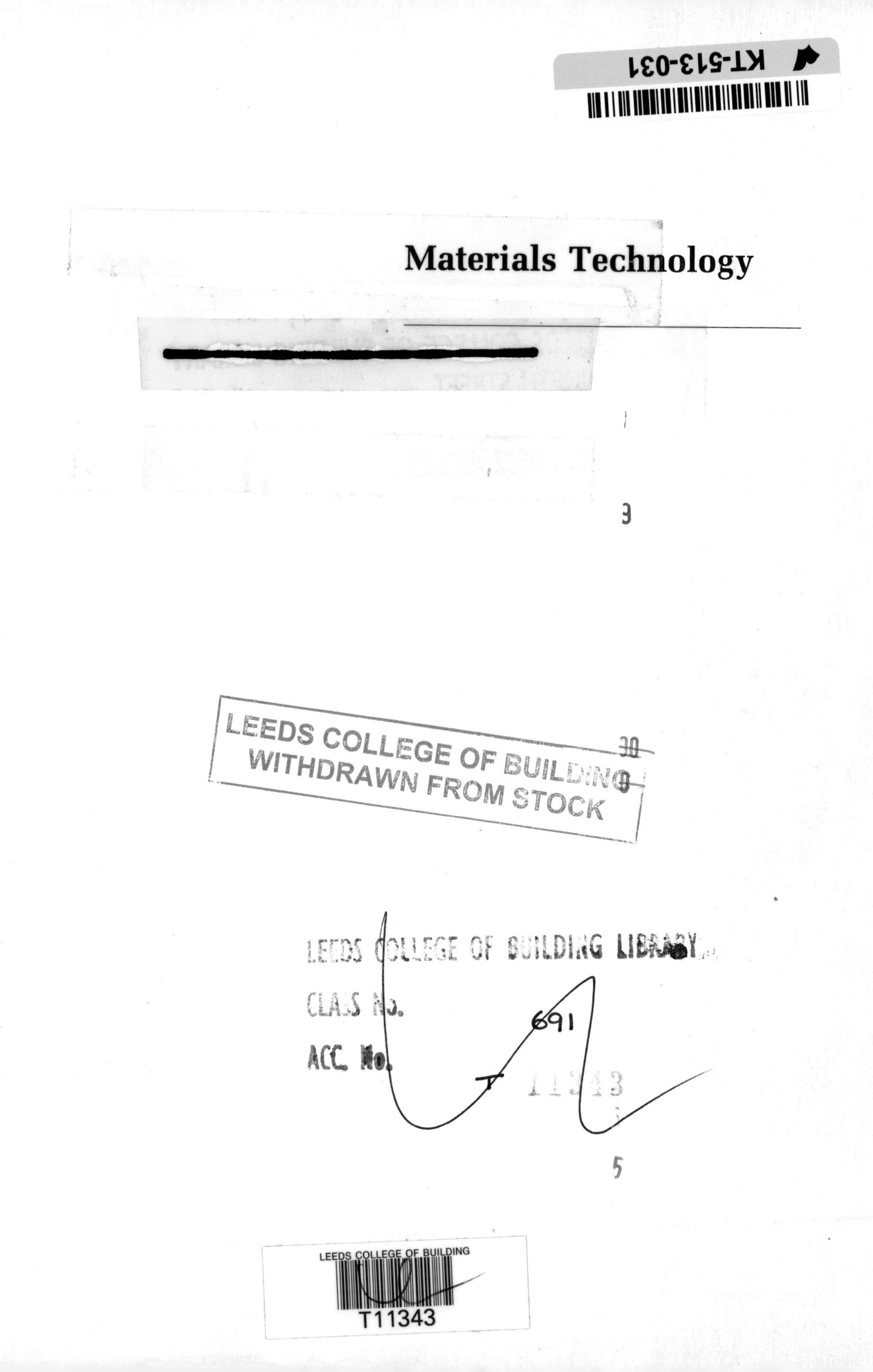

Materials Technology

MITCHELL'S BUILDING SERIES

Materials Technology

YVONNE DEAN

Longman
Edinburgh Gate, Harlow
Essex CM20 2JE, England
and Associated Companies throughout the world

First published 1996

British Library Cataloguing in Publication Data
A catalogue entry for this title is available from the British Library

ISBN 0-582-21259-6

Set by 4 in 9/11pt Times and Melior
Produced by Longman Singapore Publishers (Pte) Ltd
Printed in Singapore

Contents

Preface

Materials are the fundamental media of buildings, often dictating form, structure, quality and regional identity. As the choice and sophistication of techniques and products become wider and more complex, architects, surveyors and contractors will find decision-making increasingly demanding. By considering materials within groups that exhibit similar properties this task may be made easier; here materials are presented — for the first time in a building industry text — within their basic engineering categories of ceramics, metals, polymers and composites.

This book concentrates on the main materials used in building; for the same approach to and philosophy of the choice of finishes readers are directed to *Finishes*, also in the Mitchell's Building Series.

Since building is a visual and three-dimensional activity this book — and Mitchell's *Finishes* — include a wealth of photographs illustrating a range of examples from all over the world, not only of well-known buildings but also of those constructions that show an inventive and idiosyncratic use of materials. This book has been written to be used on many levels: as well as a source of basic information, students and practitioners alike will hopefully find this book able to inspire a more critical and better-informed use of materials. Current fields of research have been discussed to show the greater depth of knowledge that may be gained from the specialist fields of materials science and engineering.

As we approach the end of the twentieth century, issues of energy use, the choice of materials and their implications for health in the short term (on site) and long term (for building users) become increasingly more critical. The wider context reveals a need for sound and well-planned longer-term solutions. Decision-making has to embrace all of these aspects and must also now shift its emphasis from an economic basis for choice to a justification that improves the quality of the environment, is ecologically aware and minimizes damaging impact. If a wide-ranging and appropriate knowledge is applied within an ethical framework, solutions become more obvious and less difficult.

Decisions made in partnership, between people who supply and process the products and clients and users who already have their own experience and expectations, result in the best architecture; what I hope can be gleaned from this book is a new set of principles that link the technological understanding of materials with their creative use.

All information in this text is given in good faith and is the result of extensive research. However users should always verify information and check the latest standards before the application of knowledge.

Yvonne Dean
London 1995

Acknowledgements

I would like to thank the following trade associations and manufacturers who have given advice and literature:

Aluminium Extruders Association
Aluminium Federation
Aluminium Roller Products Manufacturers Association
Ash Resources (Rugby Group plc)
Building Research Association
Brick Development Association
British Cement Association
British Standards
British Steel
Copper Development Association
Eurisol UK Mineral Wool Association
European Panel Information Centre
Intermediate Technology
International Tin Research Institute
Glass Fibre Reinforced Cement Association
Mastic Asphalt Association
Pilkington
Steel Construction Institute
Lead Development Association
Ultra Polymers BASF group
Zinc Development Association and Eurozinc

Many thanks to Tohru Yamanaka, managing director of the Takenaka Corporation, in Osaka, Japan who hosted me for four days in 1990, and to all the people in the company. They showed such kindness and hospitality in looking after me and showing me the very impressive facilities and work of the corporation, ranging from the small chapel shown in this book to the Crystal Tower, 'the most intelligent building in the world', and research into dynamic solutions to balance earthquake shock in buildings.

Thanks to the following libraries and their staff: particularly Bob Cooper of the Materials and Products library; Susan and Peter Bowbeer at the main library University of North London; staff of the reference library at the Brixton Tate Library; staff at the Science Reference and Information Service, British Library, Holborn.

Particular thanks are given to Dr Roger Plank of the University of Sheffield, to Kate Ingrey, and to British Steel, since part of the text on metals was developed for the British Steel teaching pack delivered to all schools of Architecture. Thanks also to the following, all of which provided invaluable teaching material and source material for this book: the Building Centre who hosted two seminal exhibitions on tension structures and the interaction of structures and services, the Pompidou Centre who held an exhibition on Jean Prouve, the Architectural Association who held exhibitions and constructed full-size structures for the Hejduk exhibition and the Coop Himmelbau exhibition.

Lastly, this book is dedicated to the memory of seven individuals who all died tragically early, all in the period I was writing this book, all of whom I was extremely fond and miss very much:

My tutor and playmate Warren Chalk, my friend and head of department Derek Osborne, my father and severest critic Eric Dean (ex wing commander, pacifist and maker of lists), ex student and friend Ben Smith, the inspirational chair of the Architectural Association Alvin Boyarsky, my other tutor Ron Heron, and lastly Alan Blanc who sent me the most outrageous and entertaining letters.

Yvonne Dean
London 1995

The author and publishers are grateful to the following for permission to reproduce copyright material:

Cement and Concrete Association for our Table 3.8 from *Concrete practice*; Building Research Establishment for our Table 3.9 from *Design of normal concrete mixes* and our Table 10.2 from BRE Digest 69, Crown copyright; Brick

Development Association for our Tables 4.15 and 4.16 from *Improved standards of insulation in cavity walls with outer leaf of facing brickwork* and our Table 4.19 from *Designing for movement in brickwork* Design Note 10; Hodder and Stoughton for our Tables 6.6 from Higgins *Engineering metallurgy* Volume 1; Pergamon Press for our Table 7.1 from Ashby and Jones *Engineering materials*; Macmillan for our Table 9.1 from Hall *Polymer materials*; Van Nostrand Reinhold for our Table 10.1 from Hanley *Introduction to the selection of engineering materials*; Timber Research and Development Agency for our Table 11.1 from TRADA Leaflet W1 4/12 and our Table 12.12 from Timber preservation; Crown copyright.

Extracts from British Standards are reproduced with the permission of BSI. Complete copies can be obtained by post from BSI Customer Services, 389 Chiswick High Road, London W4 4AL; Telephone: 0181 996 7000.

I INTRODUCTION

1 Developing an attitude toward materials

Materials are the fundamental generators of buildings. They determine structure, overall form and quality. In the design process, the choice of materials at the earliest stage is critical and the most successful buildings are always constructed with an attitude to materials that uses knowledge to understand the full possibilities and appropriateness of choice. There are often conflicting attitudes within the building industry dependent on the role of the different professionals and their own criteria for determining choice. An engineer will have an overall ambition for an efficiency of use, and that the materials chosen will perform adequately over a period of time, which is usually the lifespan of the building. In addition, the materials chosen will often have to combine several properties to fulfil function and durability. Expectations about the longevity of the building may also relate not to the overall fabric but to how the internal fabric can cope with changing use. Buildings that are flexible have a longer life potential than those with highly specific use or users, and these expectations of use change with the economic climate and the large scale planning of buildings in cities. For example, the current economic recession in the 1990s has rapidly changed the expectations of use in building, and those buildings built in the 1960s for commercial use are seen as being viable for conversion into housing stock in a way which is far more difficult with the highly-serviced buildings of the 1970s and 1980s. Such radical change is difficult to anticipate and in future, short term solutions will be outweighed by longer term views on the equity of building stock.

There are also significant cultural attitudes which determine choice and there may be expectations as to what may be the most relevant choice. These expectations may be founded in the kind of architecture provided, its function or its permanence. There may be limitations in the availability of materials, the economics of choice or decisions which consciously use local resources. All of these initial strategies may be tempered by regional technologies in the processing of materials, their crafting, assembly or transportation.

Local expectations may be over-ridden by cross cultural solutions which take experience from other regions, whether national or international, and learning from other cultures may optimize structural efficiency or other techniques in building. The transfer of technology should always be measured carefully against the real needs of the local region, whether it is appropriate in terms of climate or regional need.

In our own late twentieth-century industrialized society we are used to thinking of buildings in terms of component technology, sometimes ignoring possible solutions that could be more traditional and labour intensive, but which could be as economic. For example, the recently completed Leicester Engineering School by Alan Short and Brian Ford (1993) is not only an energy-efficient, naturally ventilated building, but was also consciously designed to be labour intensive through its traditional masonry construction and detailing, thereby benefitting the local economy in the creation of extra job opportunities. This is in sharp contrast to the Leicester Engineering Building by Stirling and Gowan (1959) seen as pioneering engineering-led solutions using specialist skills and technologies that were current and pointed the way towards high technology solutions in building.

The economics of building decisions can then have a broader impact on decision making in materials quite apart from the longer-term traditional cost benefit analysis, which would look at facilities management, maintenance and long term equity values.

In the late twentieth century many buildings are conceived in terms of component technology, the availabilty of products and inevitably most buildings must therefore become assemblages, indicative of a production-orientated economy with an efficient manufacturing base. However, some current technologies can be as limited as traditional

Figure 1.1 Engineering School, Leicester.

methods. Some highly-engineered buildings have details and components that signify mass production methods which are illusory. The individual parts, far from showing the innovative use of standard components, display a high degree of individual craftsmanship in the making of a small number of parts, and, although the technologies available are used to their full extent, the advantage in terms of economy through mass production is lost. The use of current manufacturing technologies enables the production of these components, but they become symbols of what has been called *Neo-productivism*[1] and display the means of production rather than the ethos of the production system and its unspoken justification, which is the production of mass components. At worst this reduces a machine age aesthetic to the level of pure decoration, using non-standard components that are inappropriately engineered.

This fundamentally changes the nature of a machine age aesthetic and limits the real possibilities. The means of production now has implications in decision making through a desire to minimize energy wastage. Consequently the secondary production costs for materials as well as their primary processing energy costs become relevant.

Architecture is fundamentally a spatial discipline and the synthesis of built form relies on an awareness of choice, a sensibility to the materials used as a major determinant of structural form and as a response to local conditions and climate. There should also be a consistency about the handling of large-scale structural form as well as the resolution of detail. Within the building industry the range of understanding degrees of knowledge and practical application are wide-ranging, but for professionals to work together well and successfuly there has to be a shared knowledge base, a recognition of the overall ambition of the project and an appreciation of how the work can be organized.

1.1 Engineering categorization of materials

In this text all materials are grouped according to their molecular structure and consequent behaviour. This gives a scientific categorization related to engineering use. Materials are put into groups with similar properties and behaviour instead of earlier systems which examined materials individually in turn and continued a Victorian system of classification which became product based. As the range of materials extends in the building industry, small individual knowledge bases on each material become too cumbersome and the scientific categorization gives the student and practitioner a better understanding of behaviour and deterioration across a broad range of similar materials rather than looking for specific failure mechanisms on individual products. This helps to predict performance and aids the appreciation of the main principles that inform behaviour and choice. This strategy is also followed in *MBS: Finishes*, where the categories are explained and the actual materials are listed under applications. There is a need for practitioners in the building industry to develop a good general knowledge of materials in these categories so that initially the right choices can be made. Often this knowledge is gained by experience, especially experience of failure as a determinant in choice, which will effect careful specification. This is why there is a wide range of practical examples given in this book and illustrated photographically.

There is also a much wider range of issues which now impinge on decision making in the built environment. Knowledge of these issues brings a professional approach to the choice of materials at every level so that considerable thought goes into their selection, which can be conveyed to client and building user.

As emphasis changes in the way that architects design buildings, there are perceptible shifts in how buildings are seen and put together. Although architecture is an exercise in spatial quality, facilitated by an appropriate structure and enhanced and rooted in the materials used, the study of construction can also be too focussed on the individual elements thus reinforcing the component ideology.

Buildings may be justified on purely economic grounds but they then become mere enclosures for use rather than having any architectural or ideological significance. If that is the case they are then impoverished solutions to problems. Enclosures may not work as comfortable solutions if the feeling of enclosure is unrelated to the apparent thickness of wall. Here the material used to construct the wall will then have a *quality*, which relates to the overall idea of the space and its purpose rather than being thought of as an applied finish, and monolithic solutions may be appropriate.

There can be equally conflicting and ideological dilemmas in determining structure and approach. In structural engineering, loads can be distributed throughout a network of small components, which can each take individual stress, and failure of one of these components will not mean a catastrophic failure of the whole structure. In a monolithic building, construction (paralleled by monocoque structures in marine, aeronautical and automobile engineering) failure in one part can be catastrophic as stress will be passed directly through shells, domes and walls to affect the whole. Here ideology can be in conflict with structurally sensible solutions. Also, depending on the connections and fixings made in architecture, however many components are used, their effective connection may well make the building into a total composite that will again have to resist the stresses imposed as if the whole building were of a monolithic construction.

There are also problems in the transfer of technology and materials to situations that are inappropriate. The widespread devastation and loss of life (some 50,000 people) in the 1990 Iran earthquake is thought largely to have been due to changing systems of construction that are more westernized, and therefore inappropriate for earthquake regions. Larger-scale buildings were built in brick with discontinuity in the structure. Instead of frames, brick jack arches between steel beams were built and then infilled to give heavy floor systems which collapsed under the vibrations from earth tremors. This example illustrates the thought that is needed before making decisions that for reasons of climate or geography may be wrong. It is easy to become so comfortable with building in a particular way, using the same materials or construction, that the same methodology can be liberally and yet wrongly applied to a whole range of problems.

There is always a tendency to build and to use materials as economically as possible. Catering for unusual or extreme climatic or ground conditions means evaluating the risk of a particular event happening. The occurrence may be in cycles of 12, 50 or 100 years, but risk assessment should be used more often to establish the likely events so that provision can be made for them. The problem of building with life expectancy cycles of only 30 years means that major events can often miss a generation who will then not make provison for worst case scenarios, and this can lead to loss of life.

Examples of the usage of materials are taken from around the world. They not only show good practice and possible failure, but also emphasize state-of-the-art practice. Buildings have been chosen that are consciously well designed through the architect's or designer's own attitude to materials. The illustrations shown might seem unusual but they are intended to change the normal perspective with regard to the use of materials. As far as possible, buildings or artefacts have been chosen that show invention and also a sensible use of materials in given situations, as well as clear failure that can be an even greater learning tool.

The emphasis of this book is on the thinking of materials as a resource, with an indicator of the energy required in their processing. There is also great interest at present in the *biology* of building materials, particularly in terms of the careful choice of materials for the interiors of buildings that are healthy for users.

The description of each material looked at begins with its basic engineering classification, and works through the details of its microstructure and its behaviour at that scale, to our understanding of it at the macrostructural level. The study of the microstructure of materials gives clues as to the nature and origin of its strengths and weaknesses and its chemical behaviour, particularly in reactions with water and adjacent materials.

There has been an increase in the number of building failures and consequently litigation due to a poor application of knowledge about materials, their misuse in design and detailing, specification and workmanship. There is a need for designers and professionals to be more rigorous about their understanding, and this broader approach in the book gives an appreciation of the technical issues involved.

1.2 Detailed choice of materials

Architects are faced with decisions as to a choice of the materials they use which in turn are determined by environmental constraints. Buildings have to provide a separation between inside and outside to maintain levels of comfort, and they must be structurally sound to support their own basic self-weight and applied loadings.

In addition, there are other applied stresses in the environment which affect the durability of materials chosen which can be physical, in terms of increased wind loadings, chemical in terms of polluted environments, and there are now more complex parameters, such as electromagnetic radiation, which are not so apparent but are becoming increasingly important. Architects are major energy users through the specification of materials and components. They are now having to make choices that show they are sensible decision makers in terms of limiting the total energy demand for a building, not only by its performance but also by understanding the energy needed in the processing and transportation of materials. The knowledge base for architects to make these decisions is currently limited although there is a great deal of information now in the public domain. However, there is a lack of data that can be used as a basis for comparative evaluation and decision making.

Issues that determine choice

There is a wide range of issues that now inform decision making in the building industry and the most important ones that affect choice and building practice are outlined here. A strategy is difficult to propose as the following issues are interdependent, but awareness of these issues is the key to determining new parameters in decision making and they should include:

Storms and global warming
Electromagnetic radiation
Radon
Acid rain
Recycling and energy usage
Pollutants
Health

However, it is possible to develop a checklist which should be used as a basis for evaluating materials and components.

(1) Is the material or component energy efficient in terms of the total energy equation from primary processing, transportation and processing?
(2) Does the material or component have an environmental human impact at its source in terms of overexploitation of a resource or repercussions on a local economy?
(3) Is the material or component made from a renewable resource or from materials already in circulation?
(4) Does the material or component have potential for recycling?
(5) Is the material or component stable in terms of health and does its use or manufacture create pollutants?

Storms and global warming Since 1900 the temperature of the earth has risen by 0.5°C with a strong correlation to the buildup of carbon dioxide in the atmosphere. However, there is also natural variation of this order so the effect is not yet proved. If emissions of carbon dioxide, chlorofluorocarbons, methane and other gases continue, computer modelling suggests a predicted increase in temperature of a further degree over the next 10 years which would affect weather patterns generally, increase sea levels and change the coastline. In April 1991, Norfolk became the first county council to change its own planning policies in response to these changes. No new buildings will be given planning permission if they are within 75 m of the sea. In parts of Norfolk the coastline is receding at a rate of 1 m per annum.

If stormy weather is becoming a more frequent occurrence as a result of global warming, this will directly affect our choice of structures and materials. Negative wind pressures on buildings strip tiles and slates from roofs, and walls from buildings. This affects the design of fixing and anchoring technologies and requires stronger materials, whilst the traditional technologies rely on the weight of the roofing material to keep it in place.[2] It will also affect the design of glazing systems. It is not unusual for buildings to collapse in force eight gales.

The durability and weathering of materials are key determinants for choice in the UK where the wide range of weather conditions are far more aggressive than most climates, with a higher number of freezing and thaw cycles and generally moist conditions, putting materials under a great deal of stress. There can also be a great mismatch in the lifespan of cladding and components, with smaller-scale components lasting 10^1 years and larger-scale components and traditional building materials lasting 10^2 years. The degradation of finishes is an issue where the culture of building requires paint finishes, and these alone have an annual energy consumption of at least $3{,}275 \times 10^{12}$ joules of energy.

A joint publication from the Department of Trade and Industry and the Science and Engineering Research Council, *Industrial Materials for the Future* (HMSO 1991) emphasizes the need for addressing this aspect of materials performance, and the broad context shows that 'Repair and refurbishment already accounts for almost half of the construction activity in the UK'. Poor specification and workmanship accounts for most building failures and the amount of energy used in repair could be minimized by initial use of better materials.

Electromagnetic radiation The Riyadh stadium in Saudi Arabia is a substantial metal building and is able to pick up radio waves from a nearby radio station, and these have generated electrical currents through the building. This is an effect due to the material used and the geometry of the building, which has a number of masts and acts as an efficient radio aerial. Galvanized steel mesh has had to be introduced in a zoning pattern which extends under the ground. Ventilation ducts have had to be earthed and water supply pipes fitted with rubber connections to provide some discontinuity. Ordinary cabling systems have had to be replaced by non-conducting optical fibres.

Although this is a specific example, what is becoming apparent is the buildup of an *electronic smog* from the amount of electrical equipment in ordinary usage which can affect computers. It can be solved by placing equipment and operators in an isolating metal box or cage, sometimes referred to as a *Faraday cage*. There is no overall legislation on radio-wave interference although BS 6527 recommends limits on radiation emission, which can be exceeded by ordinary personal computers. The problem has been recognized to such an extent that Pilkingtons have been developing special glasses to act as effective screening devices. Electronic smog affects all electrical equipment, and cars are significant emitters. Radiation from radar can also affect aircraft, sometimes with catastrophic results. It is well known that the radar towers at Heathrow Airport affect the reception for local televisions. A police radio transmitter affected measurements at Jodrell Bank some 30 miles away. This was cured by lining the room housing the transmitter with aluminium foil. Electromagnetic pollution is invisible but detectable with instrumentation. Evaluation can be made of electrical devices to see if they can function where they are intended to operate under the new science of electromagnetic compatibility.[3] Electromagnetic radiation also affects people, and the measurement of radiation may become a normal part of survey work in the future so that precautions can be taken.[4] The BRE have been evaluating electromagnetic screening techniques for use in communal and civil buildings protecting sensitive equipment from interference. Shielding methods tested have used metallized glass windows and copper mesh linings in walls.[5]

Radon In 1987 public awareness of radon gas as a health hazard grew in Europe and North America.[6] Radon concentrations are found where uranium is geologically sited. There are high concentrations over igneous rocks (granite), chiefly in Devon and Cornwall,[7] but radon has also been found in limestone and sandstone areas. A national survey by the National Radiological Protection Board (NRPB)[8] also discovered high readings in parts of Somerset, the Peak District, Northamptonshire, Clwyd, West Yorkshire, Shropshire, Gloucestershire and Lincolnshire. The only areas of Britain that are completely radon free are areas of Scotland, The Borders, North East England and Lancashire. Radon decays to polonium which can be inhaled as an aerosol and is thought to be a cause of death by lung cancer for some 2,500 people annually. In America, exposure was paralleled as being greater than that suffered by uranium mine workers. Spoil heaps from tin mines in Cornwall used as aggregate in concrete or as hardcore became a source of radon gas, which then seeped up through floorboards. Remedies seem to be underfloor suction systems, which have proved successful in Sweden.

Acid rain As industry has developed since the nineteenth century, so have methods for handling pollution. Originally industrial cities were characterized by their heavy smog and pollution, as waste products, usually from burning fuel, were deposited locally. Chimneys were increased in height, and this local effect was ameliorated by a higher-level distribution of pollutants which fell over larger areas of the surrounding countryside. The height of chimneys increased further and pollutants were scattered wider afield, and today they cross national boundaries. Approximately 70% of our own emissions now leave the country and are *exported*, due to our prevailing south-west winds, to Norway and Sweden.

The chief pollutants are sulphur dioxide and nitrogen oxides mostly from power stations and cars. They convert in the atmosphere to weak sulphuric and nitric acids. Acid rain is defined as *rainfall with a pH less than 5.6*. The problem has been known for some time and monitoring since the 1960s in Sweden has shown that rainfall there was becoming more acidic, around a pH of 4.5, and they had become a dumping ground for pollutants from Britain and Europe generally. In the mid 1980s 18,000 lakes were found to be affected with about 4000 lakes virtually devoid of fish. In Sweden measures have been taken by limiting the lakes to redress the balance, but the effect is on such a great scale that it is almost a hopeless measure, and certainly not a long-term solution. The damage to Norway has been on an even greater scale. Areas similarly affected in Britain are in Scotland, the Lake District, the Peak District and North Wales. Rain falling on limestone areas can, to some extent, be neutralized by the percolation through the limestone rock. A Canadian suggestion has been to use concrete structures which would dissolve slowly over time to alleviate the problem. This would avoid fast changes in the pH of the ecosystem, but the environmental impact of such intervention seems untenable when there are realistic preventative measures at the pollution source, even to the extent of changing the systems of manufacture and processing of materials.

Acid rain falling on the soil leaches out toxic aluminium which affects the growth of trees, and consequently great forests have started to die. Already in Germany vast areas of forest have died, with a prediction that there will be a 90% loss by the beginning of the next century. As trees die, mountainous slopes destabilize and become a cause of avalanches. In 1974 acid rain fell in Pitlochry in Scotland and was recorded at a pH level of 2.4; the highest acidic level for rain ever known. Black *snow* has fallen which on melting runs into streams and there are changes in soil chemistry which release more toxic compounds affecting fish eggs and newly-hatched fish. Acid rain affects buildings directly by the dissolution of sedimentary stones, and extensive restoration is sometimes needed. It is estimated that some stonework in St Paul's Cathedral is decaying at a rate of 25 mm every hundred years; 44 flying buttresses rebuilt only 90 years ago now need replacing. Damage from acid rain ranges from staining to loss of definition on stone carvings and structural weakening.

EC directives aim to reduce sulphur emissions by 30% by 1993. If demand for energy-efficient buildings is met then this will help further to reduce demand from coalfired power stations, which are the worst emitters of sulphur dioxide. An additional solution is the use of waste flue gases, processing them through a limestone solution to produce calcium sulphite and calcium sulphate or gypsum. This can then be used to manufacture gypsum wallboard. Two new power stations planned at Fawley in Hampshire and West Burton in Nottinghamshire will be capable of producing 500,000 tonnes of gypsum annually using this process. This by-product of flue gas desulphurization, which is being fitted to existing stations as well, is expected to produce two million tonnes of gypsum by the end of the 1990s.

Even fairly remote sites of archaeological interest are vulnerable. Ancient Mayan ruins on the Yucatan peninsula are affected by acid rain as a result of pollution from uncapped oil wells in the Gulf of Mexico. Stone is being lost here at a rate of 1 mm every 10 years and there are tar-like deposits on its surface.

Recycling and energy usage In the interests of reducing carbon dioxide emissions and the saving of energy in processing primary materials, recycling is an obvious objective. Although there are collection systems for various materials, especially from household waste, there is little organized recycling within the building industry.

The recycling of constructional bulk materials is viable as the waste stream from building activities often has just four or five major constitutents and is much easier to sort than household waste.

In Britain there is greater emphasis on encouraging the recyling of materials rather than using legislation to change practice. The quantity of bulk materials discarded every year that could be recycled is phenonemal. Approximate estimates show that 60 million tonnes may be available for recycling in the USA compared with 50 million tonnes in the EC and 10–12 million tonnes in Japan. These figures are meant to triple by the year 2020.

In Britain there is an estimated 24 million tonnes of demolition and construction waste of which 11 million tonnes is being reused, chiefly in urban areas for hardcore and fill. Only one million tonnes is recycled as graded aggregate. The total amount reused still only represents 4% of material extracted as large-scale aggregates every year, estimated at 270 million tonnes of large-scale aggregate in 1989. This demand is expected to rise to 420–490 million tonnes by 2011. The repercussions of recycling also affect the use of land. One hectare of land is saved for every 100,000 tonnes of primary aggregate substituted by recycling; an additional 1.25 hectares of land is saved as the tipping of waste is avoided, not just from materials that are not recycled but also from waste at the point of mineral extraction. The mineral resources for bulk building material are not finite and waste materials in the future will be seen as a viable resource.[9]

There are general elements from municipal solid waste and clay and mineral waste which can be reprocessed by drying, gasification, oxidation and vitrification to make lightweight aggregates for the building industry. Plants using municipal solid waste and clay and liquid waste have been operating in Brisbane, Australia since 1988 and have now been built in Michigan, USA. The mineral clay wastes use material from extraction proceses that may include washed waste from coal mining, wastes from shale oil, de-watered slimes from sand washing and the overburden material from coal mining and open cast mining in general. The solid waste material consists of domestic and community waste, including sewage effluent. The building industry may well be sufficiently influenced in the future by seeing what recycled products are on offer before using primary resources. Proposed methods of control include the possibility of taxing primary materials and these will lead inevitably to a growth in demand for recycled materials. There are also very large amounts of waste materials from mineral processing. It is estimated that there are approximately 400–500 million tonnes of slate waste stockpiled, and this is being added to at the rate of 6 million tonnes per annum. There is only one quarry, at Penrhyn, where this waste is being recycled and just 0.5 million tonnes of a type 1 aggregate is being produced for local use.

Very little work has been done on the quality of recycled material. From research in India,[10] concrete made from recycled waste is more porous and requires 8–10% more water in the mix. It has a compressive strength which is approximately 10% lower with a 10–40% reduction in the modulus of elasticity and 0–20% reduction in tensile/flexural strength. Reductions of strength overall should be expected to be at least 40% but there are advantages; high porosity gives greater resistance to the effects of frost. Care also has to be taken with regard to the calcium sulphate content and more cement will need to be used. Concrete waste granulate can replace sand or gravel in asphalt mixes.

Recycled material used to be generally excluded from building and civil engineering contracts where specification clauses insisted on high-quality materials. These are usually guaranteed by the use of newly extracted minerals, which have grading and processing plant at the source of extraction to confirm quality. Relaxation on these standards in Holland and Germany now allows for recycling, and specification codes have also been relaxed in Canada and the USA to allow for the reuse of material. The greatest use of these materials is in large scale highway engineering projects where there is a great deal of waste in road renewal. Asphalt waste can be reycled relatively easily. In Holland old asphalt is softened in steam, a small amount of sand is added and the mineral oil bitumen can be rejuvenated with fresh bitumen. A regeneration agent, or 4–6% cement may be added to replace the binder. The bearing capacity is the same if the new material is laid at 2.5 times the original thickness.[11]

Car manufacturers are now forming policies which identify material components that can be recycled, even to the extent of labelling components so that their exact composition is known. Materials used to make components, such as bumpers in cars, are now being limited to one polymer, for example, instead of up to seven. Some polyethylene (PE) bottles in Holland already have standard mixtures that allow for the same kind of efficient recycling expected from glass. Proctor and Gamble are now producing Ariel Liquid bottles which use a minimum of 25% recycled polyethylene, taking four million bottles out of the national waste stream. It seems obvious that the same policies could apply to building products and pressure could be put on the industry by the specification of components that have a minimum achievement level in terms of using recycled material in their processing, or are labelled to assist long term recirculation. Specifiers should note that thermoplastics are recyclable and cross-linked polymers, which generally fall into the thermoset category, are not reclaimable. Plastic-coated materials are also difficult to recycle and the plastics element is lost forever. This philosophy is now extending to the building industry and in Germany there is an objective to ensure that at least 60% of buildings have the ability to be recycled with minimal reprocessing. In cars the non-ferrous content can be categorized into 10% stainless steel, 15% brass and copper, 25% zinc and 50% aluminium.

With regard to metals, recycling can reduce energy demand in processing. For example, the recycling of aluminium is important as it can cost up to 20 times more to process aluminium from its raw material, ore bauxite, instead of remelting aluminium scrap.[12]

There is increasing awareness, publicity and education which encourages decision making based on *green consumerism*, but there is little formal policy which is government directed as to sensible decision making. The building industry is geared to established trade practice, and specification means a fundamental rethink on the part of the professional with inadequate knowledge for decision making. There is a need to weigh choice against possible repercussions of choice and it is becoming more important to develop policies as requests are increasingly coming directly from concerned clients rather than through legislative bodies. Most guides are directed at public consumerism and many consumer *green-banded* products are suspect in terms of marketing aimed to exploit concern rather than making genuine contributions towards green policies. Some organizations such as Friends of the Earth and organizations like the Ecology Design Association publish useful information.

Recycling programmes now currently look at volume waste and try to prioritize materials that are economic to recycle and do not involve expensive techniques. There is a growing problem in the stockpiling of discarded rubber tyres (240 million per annum in the USA) which can present a real hazard if set on fire, producing toxic elements that can affect local wildlife. They can be processed physically to provide rubber crumb for new tyres and also for the surfacing of sports grounds. Rubber can also be altered chemically to allow it re-bond with new polymer material, especially polyurethane, to produce wheels and conveyer belts with good frictional resistance, and possibly gasket components in the automobile and building industry. Tyres present a vast resource given the technology to reprocess them effectively. One landfill operator now has a shredder machine for tyres which is powered by methane gas from decomposing landfill.

Many plastics used to be dumped, becoming part of landfill sites for waste, 7% of domestic waste by weight being made up of plastics. As retrieval and processing methods improve, recycling is becoming more viable.

Plastics can be recycled to produce new bottles, planks, crates, moulded chairs and rigid polyurethane insulation. Sorting is a problem which can be labour intensive, but materials produced can still make large savings and produce artifacts at sometimes one fifth of the cost to produce the original material. The sorting of plastics can recover identifiable streams of material that can be reused. Waste from cable can be recycled for the same use and PVC, when combined with vinyl compounds, can be used to form injection moulded articles such as traffic cones (which have to meet fairly high specifications) or even window frames. Manufacturers are also becoming more adept at exercising control within their own plants. The new energy management standard, BS 7750, goes some way towards encouraging a review of manufacturing process and products and processes can be kitemarked to demonstrate this change of attitude.

Dupont managed to reduce by 20% waste normally generated by manufacturing, which amounted to 500,000 tonnes of polymer a year. Dupont and Waste Management Limited are also now controlling a joint recycling programme in the USA which aims to recover some 100,000 tonnes by 1994.

As waste builds up and becomes a significant part of the infrastructure to be coped with in society, the methods for reclamation become organized on a scale that demands the organized secondary *mining* of waste dumps. This is labour and energy intensive as a process and could be pre-empted by the more organized pre-sorting and classification of materials. Although this activity is increasing, to keep its impact on the environment in perspective, only 4% of oil is used as the raw material for plastics manufacture whilst 93% of oil is still used for combustion, which emphasizes the importance of energy efficient methods for saving fuel.

Pollutants A number of the materials used in building can emit over a long period compounds that can affect people. (See *MBS*: *Finishes*.) With inadequate air changes or poorly functioning air handling systems, pollutants can build up to a point where they can significantly affect health. *Sick building syndrome* is a term in common use which encompasses problems in the quality of the environment. Pollutants are part of our way of life and we inhale compounds resulting from smoking, car exhaust fumes, and dry cleaning in daily life. Various materials *gas off*, particularly adhesives used in the manufacture of composite boards (plywoods, chipboards and fibreboards) and also ordinary paint, cleansers, propellants, various plastics and insulation materials. To minimize the level of airborne pollutants in the home and working environment, materials should be chosen that do not continuously release these substances.

Ozone is a pollutant at ground level and is generated from car exhausts, the evaporation of solvents in the paint industry, general chemical manufacturing and petrol refineries. All of these generate organic compounds that react with nitrogen oxides in sunlight. The effects of ozone in the air can irritate the lungs and throat causing coughing and chest pain. (See *Hazardous building materials*, S R Curwell and C G March, Spon, 1986.)

Health Chapters 3, 5 and 10 of this book have an addendum on health in those area where a knowledge is needed to understand repercussions of specifying certain materials and where general practice with regard to their use on site should be carefully monitored. The chief materials to monitor and the process of use is set out in *The control of substances hazardous to health in the construction industry* published by the Health and Safety Commission. Health surveillance draws particular attention to the following:

- Exposure to substances of known systemic toxicity, e.g. organophosphorous insecticide or wood preservatives.
- Asthma-causing substances, e.g. isocyanates.
- Substances known to cause severe dermatitis, e.g. epoxy resins, cement, pesticides.
- Substances which may cause skin cancer, e.g. coal tar, or pitch.
- Appendix 2 of this document gives a summary of hazardous substances found in the construction industry.

It is more difficult to provide good working conditions on site than in a controlled workshop situation, and this puts workers at a greater risk. Professionals in the industry should be aware of the risks and actively help to monitor the control of hazardous substances on site and specify materials which minimize these hazards at source. Although there may be broad clauses in contracts and bills of quantities which draw a contractor's attention to health and safety procedures, this should be reinforced by the awareness of the whole of the building team as to the risks to which people are put by the materials they specify.

The London Hazards Centre is a good source of information and has two relevant publications: *Toxic treatments: Wood preservative hazards at work and in the home* and *Sick building syndrome: Causes, effects and control*.

There is greater awareness in countries such as Germany that the chemistry of many products should be taken into account when preparing specifications for building (see *MBS: Finishes* Chapter 6). When buildings become completely sealed systems, often in the interests of eliminating draughts and high air infiltration to reduce energy demand, the air change rate drops to a level where there can be a buildup of pollutants. In these environments, especially those that provide super insulated environments, there is a greater need to specify materials and finishes that will not produce toxic pollutants. The chemistry of components can also be interactive with poor lighting design as well as poor air flow management, eventually causing *sick building syndrome*. It has been pointed out that complaints were not taken seriously by managers as their environments were often different, with higher grade natural materials such as wool carpets and timber desks, and they were often located in perimeter offices which had some external ventilation and the benefit of natural daylight. Lower grade specifications with a heavy reliance on composite materials, and desks in deep plan situations coupled with poor fluorescent lighting, would be the norm for other office workers.

Checklists published in the *Sick building syndrome* by the London Hazards Centre, appendices, give:

(1) A questionnaire which evaluates the working environment and symptoms experienced by workers.
(2) A checklist for inspecting a building which itemizes possible problems under the headings of:
   — Toxic fumes, gases or dust
   — Ventilation and fresh air systems

— Fresh air requirements
— Humidification
— Comfort
— Cleaning and maintenance
(3) Cleaning and maintenance of airconditioning systems.
(4) Design of new air conditioning systems.
(5) Sick building syndrome and the law which gives statutory guidelines for employers.

A list of consultants and laboratories which can help in specialist analysis of problematic buildings is also given.

From January 1993 an EC directive will give guidelines on minimum health and safety requirements in the workplace. Emphasis will be placed on the proper maintenance of artificial ventilation systems.

1.3 Specification of building materials

Specification is a vital part of the building process and the communication of type and quality of building materials used in construction a cause for concern. A BRE investigation between 1978 and 1983 proved that the overall quality of work related directly to the level of information support received on site.[13] It also proved that the level of management of a job improved with this kind of rigorous coordination. There was also less conflict between consultants. These conclusions were backed up by a separate study undertaken by the York Institute of Architectural Studies.[14] This concentrated on how architects organized the flow of information, responded to their work load, and how projects were affected through the poor collection and organization of information, in a working environment that inevitably fragmented the working day.

Co-ordinated project information was a system set up in 1979 between several main bodies in the construction industry which included the RIBA, RICS, BEC and ACE. It sought to relate properly all the documentation used in building projects in drawings, specifications and bills of quantities. It established conventions that could relate specification clauses directly to coded drawings. The advantage of the discipline of such a system is that it ensures a proper cross checking of information and eliminates duplication and often conflicting notes that might appear on drawings. It also encourages the proper drawing up of a specification that is building specific. Drawings are related back to the National Building Specification and to the National Engineering Specification as well as to the SMM 7 standard descriptions. It was seen as not only an organizational tool but one which would minimize litigation through poor specification, and badly arranged information which caused delay and also led to defective work and poor cost control.

Co-ordinated Project Information is characterised by some 300 work sections which are arranged in 24 major areas showing directly how a job is organized and how trade sections would be split and let to subcontractors.

References

1 Kenneth Frampton 1982 Modem Architecture and the Critical Present. *Architectural Design* profile: p. 5. This presents a discourse on regionalism.
2 P S J Buller 1986 *Gale damage to buildings in the UK: An illustrated review* BRE, Watford. This publication analyses failures in building due to wind loading through design and construction faults on site.
3 Barry Fox 1989 Electronic smog fouls the ether. *New Scientist* 7th April: pp. 34–8.
4 Cyril Smith and Simon Best 1989 *Electromagnetic man: Health and hazard in the electrical environment* Dent.
5 BRE *Annual Review* 1991.
6 Fred Pearce 1987 A deadly gas under the floorboards. *New Scientist* 5th February: pp. 33–5.
7 News item 1990. *New Scientist* 27th January: p. 30.
8 General information and advice, National Radiological Protection Board.
9 *Occurrence and utilisation of mineral and construction wastes* 1991 DOE. This publication analysed the origins of builders' waste in Germany and showed that 92% came from modernizing houses, 5% from new build and only 3% from demolition.
10 N K Bairagi, K Kavande and V K Pareek 1993 Behaviour of concrete with different properties of natural and recycled aggregate. *Resources, Conservation and Recycling* August.
11 Charles F Hendrik 1986 Reuse of road-building materials. *Recycling International.*
12 John Newell 1990 Recycling Britain. *New Scientist* 8th September: pp. 46–9; and Julie Johnson 1990 Waste that no one wants. *New Scientist* 8th September: pp. 50–55.
13 *Quality control on building sites* BRE Current Paper 7/81.
14 *Design decision-making in architectural practice* IAAS Research Paper 19 1982.

Additional references

Environmental Information: A Guide to sources, Nigel Lees and Helen Woolston, The British Library, London, 1992.

Energy efficiency and the environment, OECD/IEA, Paris, 1991.

The 1993 environment catalogue, HMSO Books, 1993.
Energy efficiency, Ian Christie and Neil Ritchie (Eds.), Policy Studies Institute, 1992.
European directory of energy efficient building 1993, James and James Science publishers, 1993.
Elsevier materials selector, Norman A Waterman and Michael F Ashby (Eds.), Elsevier Science Publishers Limited, 1991. (Note: This is a major reference work of three volumes containing details on all aspects of materials, their relative properties and gives parameters for their choice.)

II CERAMICS

2 Introduction

The range of materials in this category form the greatest bulk tonnage used by the construction industry. The minerals extracted are used to make the bricks, other traditional ceramics such as tiles as well as the cementious component of concrete and concrete blocks. The implication of using and specifying these materials in the building industry has a major impact on the landscape and considerable energy implications in the heavy plant used for extraction, fuel for transportation and energy in processing. The energy used in processing is mostly used in bringing kilns up to temperature; very little of the energy is used in the reaction processes to transform the minerals.

In the UK alone annual figures in the early 1980s showed that over 100 million tonnes of sand and gravel was being extracted for making concrete with a further 30 million tonnes of rock for coarse aggregates and stone. The making of cement used 100 million tonnes of shale and limestone. The extraction of minerals peaked in the 1970s to meet demand in the construction industry and the resources are not finite. Manufacturers who set up plant for making concrete blocks near a local resource for aggregate are having to transport aggregates of the right grade from further afield. As extraction processes become more visible and major landscape features are altered in a way that becomes unacceptable planning processes are tightened. Mineral extraction changes the profile of land by altering physical features, by creating spoil heaps and large pits which cannot always be filled or flooded, and land passes into a category of being unusable and termed derelict meaning:

> . . . land so damaged by industrial or other development that it is incapable of beneficial use without treatment.[1]

This category relates to approximately 50,000 hectares of land in mainland Britain. The full impact of mining is only ever visible to a minority. One example in Malta which shows the extreme results of mining is the complete alteration of the south-west corner of the island which has lost virtually all of its original landscape. A building boom in the last twenty years has used stone blocks cut directly from limestone beds on the island, leaving a multitude of square pits, highly dramatic from the air and yet hardly visible from the surrounding roads.

Although there is a growing use of waste materials from major producers in industry they still are relatively small in tonnage terms. For example, in 1980 9 million tonnes of slag from blast-furnace waste was mostly used as aggregate in concrete and road building, with a further 6 million tonnes of pulverized fuel ash from power stations for lightweight aggregates and cement making. Ceramics are by far the largest group of materials used in building. They range from the traditional ceramics of bricks, tiles and stone, to the composite materials of concretes and cements, and the more advanced engineering ceramics used in car engines and as cutting tools, with a limited application at present in building. Although glass ceramics can now be made with a guaranteed zero expansion they are only used for optical mirrors for astronomy or for parts in space vehicles.

Ceramics are broadly non-metallic materials, and not organic, with high melting points. They are usually oxides, carbides, nitrides, borides and silicides. As they are chemically inert they have the advantage of being inherently corrosion resistant in our atmosphere and can be used as the finished material. They are weather resistant, and high hardness, low density ceramics have been developed as military armour but could have applications as longterm durable components for cladding and services in building. Currently their high cost makes their use prohibitive.

Ceramics have problems of low fracture toughness and are sensitive to high stresses if applied locally. They are non-ductile and a repercussion of defects leads to notch sensitivity and the propagation of cracks, giving a characteristic behaviour often described as brittle. Strength is dictated by the way in which defects can be controlled, increasing material fracture toughness. If this is optimized, then ceramics can be engineered to give the same performance as metals in terms of non-catastrophic failures. In order to optimize the behaviour of ceramics it is becoming more common to reinforce them with fibres. A macro scale ceramic composite of this nature is reinforced concrete and, at a smaller scale, glass-reinforced cement. To optimize the behaviour using fibres it is more effective to have a weak bond in between matrix and fibre. This means that stress is more difficult to transfer and the stress applied is dissipated. If the fibre–matrix bond is strong, cracks can be transferred directly. Bonding mechanisms that enable the fibre to move or de-bond under stress can toughen the composite to withstand stress 500 times that of a monolithic ceramic. The smaller the fibre diameter, the greater the number of fibres needed to reinforce the matrix. However, a greater area of interface is then achieved between matrix and fibre with an increase in the toughening mechanism provided. Glass fibres are still one of the strongest fibres produced although in advanced engineering ceramics these are now being replaced by silicon carbide whiskers, or silicon carbide or alumina fibres.[2]

2.1 Advanced ceramics

Most advanced ceramics are characterized by being fine grained and low porosity materials, processed to remove as many defects as possible.

These ceramics group into three main categories which are differentiated by the way in which they have been reinforced and formed. They are:

Fibre type These can be polycrystalline, meaning they have been reinforced with continuous or chopped fibres, or single crystals (more commonly referred to as whiskers).

Matrix forming These ceramics include powder processing which may use powders that are sub-micron in their dimensions, compacted with organic binders then sintered, or used as liquid slurries in slip casting, or processed in a gaseous phase using chemical vapour deposition.

Consolidation These use either shrinkage methods in fibres or fill an open fibre-network with a fluid matrix which is then hardened.

Although the *density is strength* principle might be followed for most ceramics, there are occasions when porous ceramics need to be manufactured for lightweight, high-strength materials or as insulation. Sometimes a matrix of fibres might be coated with a slurry which is then extracted by vacuum techniques leaving a rigid fibre skeleton.

The physical properties of some common compounds in engineering ceramics are given in Table 2.1.

2.2 Definition of a ceramic

The main ceramic materials covered in this section include concrete, brick and blocks, stone and glass.

Ceramics cover a range of materials that all have a similar range of properties and behaviour. This widens the traditional classification of ceramics to cover materials that vary in their origin and appearance. Ceramics can cover natural as well as man-made materials, i.e. minerals, rocks, cement gels, glasses and sintered clays which give us our traditional ceramics. These traditional ceramics are derived from the mining of clays which are fired to give pottery, bricks and tiles, and are a combination of flint, feldspar and clay. It is important that ceramics are properly understood as covering a wide-ranging group of materials. They are all:

Inorganic One exception is diamond-carbon which is sometimes classified as a ceramic.

Non-metallic properties But metallic elements may be present.

Ionically and covalently bonded

Combination of metallic/oxygen atoms More generally gaseous and metallic elements, e.g nitrogen can be substituted for oxygen in many aluminosilicates which are then called *sialons*.

They often share a range of properties which include:

Hardness

Thermal insulators Good fire resistance

Table 2.1 Common compounds in engineering cermanics

Compound	Toughness (K_{ic})	Hardness (Vickers)	Density (g/cc)
Zirconia ZrO_2	12	1120	5.75
Alumina Al_2O_3	5	2000	3.8
Silicon carbide SiC	4	2100	3.1
Silicon nitride Si_3N_4	6	1600	3.2

Electrical insulators Density is important here for effective insulation and temperature is now relevant with regard to the development of super cooled semiconductors.
Chemically stable Sometimes described as inert within the range of weak compounds with which they may come into contact.
Dimensionally stable
High strength at high temperatures
Wear resistant
Non-ductile Brittle behaviour.
High compressive strengths
Multiphase materials

The ceramic materials used in building exhibit all these properties. The advantages of using many ceramic materials is that they are already environmentally stable and will not oxidize further in the atmosphere. Unlike metals, which generally exist in an instable state and must be protected, ceramics are unlikely to react with elements in the atmosphere and are normally used in their natural state, i.e. their body composition can be their finished skin. This makes them economic in terms of maintenance. Problems are more likely to occur when they are combined with other materials to give different properties. For example, when concrete is in combination with steel — initially to improve tensile strength by making a composite — carbonation and the subsequent corrosion of reinforcement will make the whole material unstable. Thermal expansion coefficients also need to be considered, as metals have values 10 × that of ceramics. Mechanical fixing details generally are also a source of failure. The basic ceramic material usually degrades slowly, and is a long life component, whereas the fixing mechanisms are often highly stressed and if corrosion occurs in fixing components, failure is more dramatic. The other major advantage of most ceramic components is that they are non-toxic.

Traditional ceramics

This category should be used when talking about fired clay products. Depending on the clay bodies used, the admixtures and the degree of firing, the phase microstructure of ceramics can be manipulated.

Earthenware clays will produce a soft porous material after firing at low temperatures. On fracture, the material will show cleavage around individual particles. Glazed material shows a noticeable boundary between glazed and unglazed sections. It will have a Moh's hardness of about 4 or 5.

Stoneware clays are fired at higher temperatures with a hardness of 7 or 8. If fractured, the material will break as a glassy body with a continuous fracture plane. A stoneware glaze has the appearance of fusing with the clay body, which is extremely dense and non-porous. The silica in the stoneware forms the glassy matrix in which other particles are suspended.

Porcelain is the most glass-like body, with a glaze appearing completely integral with the body. Although a fine-grained material, it is extremely tough and durable. The microstructure of porcelain reveals needle-like crystals of mullite. These act like fibre-reinforcing elements in a glassy matrix and prevent crack propagation.

Moh's scale of hardness

This is a scale of hardness that relates to the ability of a material to resist a scratch defect from a mineral. The tests should be made by hand. The examination for scratch deformation is by eye. To test a material, minerals that have been recently fractured with a sharp edge of increasing hardness are drawn over the test surface. This method of testing is described in the British Standard relating to ceramic floor and wall tiles. (See BS 6431 Part 13: 1986.) Table 2.2 shows a practical estimation using common objects if the correct minerals for testing are not to hand.

Multiphase materials

In solid materials there may be regions with differing molecular structure. In many rocks these would be seen clearly as different mineral forms with individual, identifiable characteristics. Most ceramic materials could be called *multiphase solids*, as they usually are composed of particles with their own individual structures set within a matrix which may be glassy but also has its own identifiable structure.

2.3 Bonding

Bonding is a term that has several different definitions according to which discipline is using the term. In building

Table 2.2 Moh's scale of hardness

Scale	Mineral	Practical estimation
1	Talc	Can be scratched by fingernail
2	Gypsum	Can be scratched by fingernail
3	Calcite	Can be scratched by a copper coin
4	Fluorite	Can be scratched by steel
5	Apatite	Can be scratched by steel
6	Feldspar	Scratches steel with difficulty
7	Quartz	Scratches steel with ease
8	Topaz	Scratches steel with ease
9	Corundum	Scratches steel with ease
10	Diamond	Scratches steel with ease

it relates to 'the securing of a bond between plaster and backing by physical means as opposed to mechanical keys'.[3] Mechanical keys, or mechanical bondings, refers to an interlocking of parts by their profiling, such as in dovetailing, or by using fixings such as bolts. Physical bonding refers to good absorption by the substrate with a substance and good adhesion. Adhesion varies in its definition according to the scientific discipline concerned, but in physics is generally defined as:[2]

> Intermolecular forces which hold matter together, particularly closely contiguous surfaces of surrounding media, e.g. liquid in contact with a solid.

Adhesion is often achieved by a physical attraction through Van der Waal's forces between pore walls (usually displaying *negative* charges from compounds containing oxygen) and their initial attraction with water molecules in solution (which are dipoles, acting as small magnets, with *positively* charged hydrogen ions).

References

1 Government definition given in Block 2 of the Earths Physical Resources *Constructional and other bulk materials* Open University, 1984: p. 62.
2 Robert Lundberg 1989 *Fibre-reinforced ceramic composites* Department of Inorganic Chemistry, Göteborg University.
3 *Chambers Dictionary of Science and Technology.*

3 Concrete

3.1 Definition and early usage

Concrete is an unusual material as it is an artificial composite, using raw materials and then rebuilding them into rocklike forms imitating structures known as *conglomerates*. These are particle composites of stones and other material embedded in a matrix of silt and sand that has hardened over time into a sedimentary stone.

Although this seems straightforward as a concept the resulting material of concrete is artificial and complex due to the chemistry of cements. Although the constituents may be basic in origin, their processing and mixing can be critical, affecting both final strength and durability. Early concrete making always used local materials and, in Italy, the ready supply of sands and gravels plus a binding mortar from the volcanic material pozzolana (mixed with lime and wetted) provided the right ingredients for large scale monolithic constructions.

Vitruvius wrote about the techniques of using concrete and, although these were a source of information for Renaissance authors, there is no large-scale use of concrete documented until the nineteenth century. Uses of the material were often domestic in scale and there are examples all over Europe of dwellings which use a form of earth building modified by the addition of binders, often using lime cements. Stone was considered a superior building material for public works and early concrete and earth buildings were often imitative of these buildings with expressed courses and mouldings. Alternatively, rendered finishes could be formed to provide all of the detail to imitate courses, or run with moulds to give string and cornice details. Rondelet (1812) said:

> The rich traders of Lyons have no other way of building their country houses. An outside covering of painting in fresco attended with very little expense conceals from the spectator's eye the nature of the building material and is a handsome ornament to the house. Farmers have them simply whitewashed but others who have a greater taste for ornament add pilasters, architraves, panels and decorations of various kinds.

Figure 3.1 The Coliseum in Rome. The mass walls use brickwork as both permanent formwork and large scale aggregate.

Wall building could be thinner and more controlled and,

Figure 3.2 This house in Charenton, formerly a town in its own right, and now engulfed by Paris, shows the rendered details used to imitate stonework, a technique that became more acceptable as early concrete mixes replaced traditional materials. The early techniques of mud or earth construction (used more extensively on the continent than in Britain due to soil conditions) were improved in France.

Figure 3.3 A first storey earth wall on the edge of Exmoor proved more difficult to demolish than anticipated. It had hardened over centuries and demolition workers had to take a break after making little impact on this last section with pickaxes.

instead of being built by hand, techniques were used involving the ramming of mud and earth between temporary timber formwork made from tongue and grooved boarding. To improve strength it was logical to look at the binding of earth and stones with a weak cement which led to a more logical selection of stones for aggregates. Mix design was vastly improved by Coignet, a director of a chemical factory in France, who introduced sand into the mix design and looked at the effect of water ratios on strength. Coignet had started to introduce iron rods instead of beams into his concrete construction and this was taken further by Hennebique who used reinforcement to tie beams to columns, and cranked reinforcement at junctions to counteract shear stress.

Developments in the better engineered concrete structures were only expressed in industrial buildings. Hennebique held patents for his system and tended to supervise his own approved contractors. One building in Wales, erected between 1897 and 1899 for Weaver and Company (flour millers and grain importers who wanted a fireproof building), was built using French labourers and materials, and was reputed to have used bent or cranked reinforcement rods for the first time. It incorporated a cantilever section of 4.26 m, and the flat roof also acted as a reservoir for 20,000 gallons of water. It took just seven months to build, and survived for a hundred years, only being demolished in the late 1980s.

Concrete has a long history and has developed from the use of roughly combined local aggregates and cements to precisely measured quantities of materials which have a known composition, strength and purity. The greatest variant is the cement.

In terms of cost it is worth mentioning that the pumping and placing of the concrete can account for 11–13%, the material constituents 14–16%, reinforcement accounts for 25–30% and the formwork varies between 45 and 50%. The high cost of formwork is due to the labour involved.

3.2 Portland cement

Portland cement used today as the binding agent or matrix in mix design has straightforward origins, resulting from limestone and clay which is heated to a clinker and then ground to a fine powder. This is then mixed with a small proportion of gypsum (calcium sulphate) which affects the

Figure 3.4 An unusual ad hoc approach to the use of concrete in a pill box in Northumberland. Hessian bags have been filled with wet concrete and then 'coursed' to provide a wall.

Figure 3.5 The hessian has rotted away leaving the seam details of the original bags. This shows the accuracy possible in reproducing detail in concrete, and its durability.

rate of set of the material. As all water has been drained off from the original material, the resulting powder is unstable and will readily recombine with water to form a hard solid. When water is added to the cement mixture it initiates the chemical reaction of *hydration* or absorption of water, and the resulting material resembles a mineral form known as *tobermite* with a formula of:

$$Ca_3Si_2O_7H_2O$$

This is a complex silicate, sometimes described as a series of linear chains of amorphous silicate, which eventually form a rigid, sponge-like material.

The actual chemistry of the reaction of Portland cements is then very similar to the tobermite mineral, producing complex calcium silicates and aluminates, but it is not exactly the same and the reactions are still not fully understood, nor are they entirely predictable as there are so many constituents and variables. The hydration reaction explains why cement powder has to be kept dry on site. Opened bags absorb moisture from the atmosphere and start the hardening process known as *air setting*. Further mixing with water will produce a *flash set* or a complete hardening very quickly.

Cements are also regarded as *gels* which are mixtures of solids and liquids. In cement the chemical reaction progresses when water is added and sheet silicate structures form, sometimes curling into fine tubes to create a three dimensional framework that holds other cement particles and water by electrostatic attraction. Every particle is wrapped in cement and connected to adjacent particles by cement *bridges*. This mass of solid material can be regarded as a complete network of suspended solid which is homogeneous throughout the mix and which does not settle.

As the reaction relies on water making intimate contact with cement particles, the finer the particles the greater the total surface area affected, with a reaction being taken to completion. Consequently control of the actual fineness of the cement is a major controlling factor in the strength of the concrete, and covered in BS 12 as an important parameter.

Cement is known as a *rigid gel*, a network of material so complex that it traps liquid permanently. Rigidity is a characteristic of a material having a structure which cannot distort and soften after the application of heat. There are *nonrigid gels* or *elastic gels* and one example of these is asphalt. In rigid or elastic gels the aim is to produce a dense mix that eliminates voids and to achieve greater strength as a result. Concrete uses aggregates as packing material to minimize the use of expensive cement as a binder.

Even though concrete is a rigid gel, the silicate sheets formed have the ability to slide over each other and this happens when concrete creeps under stress. This kind of deformation can recover elastically, returning to its original shape when the stress is removed.

Once initiated the chemical reaction, or the hardening mechanism, is a continuing process as more cement particles convert and this explains why the strength of concrete can increase from the initial *set* over time. This

Figure 3.6 Housing by Jourda and Perraudin, 1982–84 Isle d'Abeau showing the technique of using earth in large rammed earth panels with no protective render, the use of which is thought to accelerate decay rather than act as a protective element in this area.

is reflected in the advice given as to when to dismantle formwork holding the initial wet material, with an agreed time for testing all specimens for comparable results. Concrete will carry on increasing in strength and will reach maturity after about 20 years. The crushing strength can almost double in that time. For the purposes of grading and testing concrete, the term *ultimate strength* means strength after 90 days, and the time for testing at 28 days is when the concrete has reached 80% of its maximum strength.

As the hydration process is an exothermic chemical reaction, a great deal of heat is generated. Although this presents few problems in temperate climates, it can lead to thermal cracking due to an initial reaction with one major constituent of the final gel, *tricalcium aluminate*, which can make up approximately 11% of the total concrete weight. In major engineering structures with a great mass of concrete, such as the Hoover Dam, the concrete mix has to be altered to prevent the effects of this compound. In these large civil engineering structures the heat generated can be so great that often pipes are laid in the structure and pumped with refrigerant until the concrete finally sets. They are then left in place as permanent reinforcement.

In BS 5328 *Methods for specifying concrete including ready mixed concrete* the definitions are given for cement in 3.3 as a hydraulic binder comprising of three options which are:

- Hydraulic cement-active hydraulic binder formed by grinding clinker and complying with BS 12, BS 1370 or BS 4027.
- Hydraulic binder-controlled process with Portland cement combined in proportion with a latent hydraulic binder of pulverized fuel ash or granulated blast furnace slag complying with BS 6588, BS 146 or BS 4246 (depending on the binder used).
- Hydraulic binder of Portland cement to BS 12 with pulverized fuel ash (PFA) to BS 3892 or granulated blast-furnace slag complying with a British Standard in the course of preparation and complying with BS 6588, BS 146 or BS 4246.

Main compounds formed in Portland cement

There are four important compounds which have been isolated in cement and which have specific properties with regard to their setting and hardening. By designing the cement to alter the proportion of these compounds formed, control is achieved over its setting and strength.

Tricalcium aluminate ($3CaO.Al_2O_3$) This compound hydrates rapidly with great heat and although it assists in the initial setting of cement it contributes least to its final strength. It is vulnerable to chemical attack particularly to sulphates from groundwater which can initiate cracking if not properly controlled.

Tricalcium silicate ($3CaO.SiO_2$) This gels very quickly with great heat determining the initial early strength of the cement within the first 14 days of setting.

Dicalcium silicate ($2CaO.SiO_2$) This compound forms slowly with a comparable degree of heat evolution and progresses the strength of the cement after the 14-day period to the 28-day standard for testing and beyond. It has greater resistance to chemical attack, shows greater durability and low drying shrinkage.

Tetracalcium aluminoferrite ($4CaO.Al_2O_3Fe_2O_3$) This compound makes no significant contribution to the cement and could be regarded as a product of the reaction, taking space as a filler.

Water:cement ratio

In order to achieve control over the strength of the concrete, the elimination of voids is paramount, and this means having control over the water content. Water can be categorized as interacting in three ways in concrete.

(1) It combines chemically with cement in the newly-formed hydrated compounds. This is about 25% by weight of the fully hydrated cement and cannot be evaporated.
(2) It is held in the gel pores by electrostatic attraction and is strongly bound to the pore wall surfaces, amounting to about 15% by weight of cement content.
(3) It is water not needed in the chemical reaction and is unable to reach the saturated walls. It is regarded as excess water in the capillary walls and if there is an excess of water in mixing it will enlarge the mix. As the hardening process takes place this additional water is able to evaporate out of the mix, leaving voids.

This is why keeping to the correct water:cement ratio is so important. Excess water will lead to greater porosity, low density and consequently lower strength. Using the ratios of (1) and (2) combined, only water amounting to 40% of the weight of the cement is needed. If the water content is increased for the sake of workability to 60% by weight of the cement, this can lead to halving the strength of the concrete. The water:cement ratio is normally shown as:

$$\frac{\text{total weight of water}}{\text{weight of cement}}$$

or sometimes referred to as the *free water:cement ratio. Free water* means the total weight of water in the concrete after allowing for absorption by the aggregates. In order to reach the correct water content, the mix design must allow for this level of absorbency otherwise there will be an inadequate amount of water for the reaction.

The quality of water must maintain a purity equivalent to drinking water and is normally specified to BS 3148. Sea water cannot be used, not for reasons of strength but due to the action of free chlorine ions that will initiate corrosion of steel reinforcement. Water from streams that has suspended organic matter can affect the hydration of the concrete, its strength and its long-term durability. The reactions that form cement are complex and any impurities or additional compounds will affect the mix.

3.3 Earth as a building material

In some regions stone as a major building material for walls was not available and there was a well developed worldwide tradition of building with earth or *pise* as it is called in France — often having a history which might extend back 2000 years. François Cointeraux was generally recognized as the greatest user and innovator of this material in France and used traditions that had resulted from this long-term experimentation. However, he adapted the technology of building in this material to buildings that were greater in typology than the rural vernacular of dwellings, extending the technology to public buildings, schools and factories. As a material it was available to all and there were political reasons for its use which were a physical expression of the idealism of the French revolution. There are very specific areas throughout Europe where the soil is suitable for this type of contruction. Interest was expressed in the twentieth century by architects who learnt from their local vernacular, Le Corbusier from the Lyon and Grenoble regions, Frank Lloyd Wright from the adobe architecture of the early Americas, Clough William Ellis from cob and stabilized earth techniques used in England, and Hasan Fathy from the mud brick tradition used in Egypt. Modern use of earth building has been developed in America particularly in New Mexico and in Australia, Later use in the twentieth century in Europe has been more limited, but the experimental Domaine de La Terre built in the new town of Isle D'Abeau in France showed that use of these techniques is still relevant today and it should be considered as a serious option in building. As a local material it immediately has advantages through eliminating large processing and transportation costs and any cost transfer is to the support of local labour. The main reason these techniques are not used today is due to the ease with which all countries have adopted Western industrialized techniques and the assumption that these are the only *correct* methods of building even though over one third of the world's housing stock uses earth as a building material.

There are some areas in the world where building in earth is still part of current building practice; in America, Africa and Australia. In these areas a local tradition is reinforced with a continuing interest by professionals and also acceptance by authorities within the terms of local building codes. The difficulty of building in earth should not be underestimated as the method of earth building is always highly specific to a locale. Ways of working the soil, the choice of binder, the method of compacting earth, whether into bricks or large blocks or tamped into walls between timber formwork depend on the local soil condition and techniques and material available locally. Even the method of compacting earth is critical and over-compaction with modern mechanical plant can lead to unnecessary fracture of the material. Earth construction as practised successfully was communicated orally and expertise is lost as craftmen die without recording local mixes and techniques. This means successful building practice is lost which was based on an empirical method of building through trial and experimentation over many hundreds of years. Methods of using earth construction today should seek to optimize the local soil condition and resources available and accurate soil analysis is fundamental.

There are few centres of expertise and the group CRATerre, based in Grenoble, act as an information gathering centre and have published useful working guides based on practice.[1] Because of the long and substantial history of earth building in the region the Laboratoire Geomateriaux at ENTP (École Nationale de Travaux Publics)[2] provides consultancy advice and has a list of publications.

3.4 Types of cement

Ordinary Portland cement (OPC)

This is the most common cement used amounting to 90% of the UK production. It is manufactured to BS 12. Improvements have given higher strengths although these are accompanied by greater early heat of hydration. This gives tolerance to higher water:cement ratios, although long-term durability, which relies on density of the mix, may be reduced.

Rapid hardening Portland cement (RHPC)

Although these cements conform to BS 12 the only difference is that rapid hardening cement has a finer particle size. This gives greater surface area for the hydration reaction to take place and increases the rate for gaining strength.

Advantages may be found in precast concrete production allowing for the earlier striking of formwork. The strength increases faster than OPC, which could still be affected by poor site conditions.

Sulphate resisting Portland cement (SRPC)

Sulphates with water will cause ordinary Portland cement to deteriorate, affecting concrete and mortar mixes. A sulphate-resisting cement is commonly specified for foundations or slabs when ground conditions show the presence of sulphates. Sulphates can also be found in some bricks and chimney flues as a combustion product. If sea water is present or repairs need to be made to concrete or brickwork where sulphate compounds have been detected, SRPC complying to BS 4027 should be used. This cement has a lower content of tricalcium aluminate which is the most affected cement compound in this situation. Iron oxide is added to the mix and more of the compound tetracalcium aluminoferrite is formed. The mix design should be controlled to ensure adequate density and strength, depending on the local concentration of sulphates. (See BRE Digest 250 *Concrete in sulphate bearing soils and ground waters*.) Do not use SRPC in combination with pulverized fuel ash cement if specified to BS4027. There may be exposed details from slab or column bases which should take account of this cement setting to a darker colour. See BS 8110 and BRE Digest 250 for advice generally about the specification of cement in these situations with regard to percentage content and impermeability.

White Portland cement

This cement complies with BS 12 but has a low iron content and incorporates white china clay and white limestone as the alumina and calcium carbonate components, respectively. Although it is used because of its colour, care should be taken in ensuring that batches of material have consistency in colour. An alternative solution is to use grey Portland cement and add white pigment. This is used as a technique by some precast concrete manufacturers for greater control over possible colour variations.

Low heat Portland cement

This cement is only used in large engineering structures where the amount of heat generated during hydration is sufficient to cause cracking. It is not a standard product and is manufactured to order to BS 1370. In certain situations the use of sulphate-resisting cement may be appropriate as it also has a lower rate of heat evolution than Portland cement.

A cement similar to Portland pulverized fuel ash cement (see below) with a pulverized fuel ash content between 35 and 50%, produced to BS 6610 can be used as a low heat cement.

Masonry cement

Ordinary Portland cement and sand mixes are often too strong for general rendering, brickwork or blockwork mortars. Instead of adding lime to the mix, which is unpleasant to work with, masonry cements are available. These incorporate finer mineral powders and air entraining agents which make the cement more flexible. (See BS 5224 *Specification for masonry cement*.)

Portland blast-furnace cement (PBFC)

This cement, which incorporates ground granulated blast-furnace slag (g.g.b.s.), is being used to a greater extent in the industry and combines OPC with 30–35% blast furnace slag. It takes time to gain strength which minimizes cracking and is more resistant to chemical attack. The early rate of strength development is reduced although the 28-day strength is equivalent to OPC. (See BS 6699. See also BS 146 for PBFC with a ground granulated blast-furnace slag content of not more than 65% and BS 4246 for a GGBS content of between 50 and 90% used as a low heat cement.)

Portland pulverized fuel ash cement (PPFAC)

This is a mixture of Portland cement with 15–35% pulverized fuel ash. (See BS 6588.) Pulverized fuel ash is waste material from the combustion of pulverized coal in power stations. It is a glassy spherical aluminosilicate with around 50% silica content and 25% alumina. It is used in combination with Portland cement where the hydration reaction produces calcium hydroxide. The calcium hydroxide will then combine with the alkali-soluble silica and alumina to form calcium hydrates which are then cementitious compounds. Like PBFC it has a slow rate of strength development but there are some advantages. With a percentage of PFA higher than 25% there is greater resistance to sulphates, and over 30% will give additional protection against alkali silica reaction. (See also BS 3892 Part 1 for the use of pulverized fuel ash as a cementitious compound in structural concrete.) It shows that structural cements using this aggregate are restricted to the finer grades of ash classified as grade A. Grade B ashes are used for non-structural components such as paving slabs. The cement can achieve higher strengths than OPC over time and will need less water than OPC due to the particle size, shape and grading. As the hydration reaction is slower, less heat of hydration is generated and this minimizes thermal cracking. This is particularly important in large placements of concrete and there is greater demand for its use in large-scale civil engineering projects. The final cementitious product has less porosity and the reduction of permeability leads to greater durability generally for concrete. It is this property which restricts chloride ion penetration and minimizes sulphate attack. The flow characteristics of this cement are better than OPC and lead to improved pumpability and easier placement of concrete.

Resistance to freezing and thaw cycles is improved using PFA which has a better distribution of small air bubbles (less than 0.2 mm in diameter).

Water-repellant cement

This is achieved by using additives to Portland cement, often metallic soaps. It helps to close the pore structure in concrete and works by changing the electrical charge of the surface and repelling water molecules. It gives concrete surfaces greater resistance to weather staining. Manufacturer's instructions have to be followed to ensure a dense mix. Waterproofing systems should be checked carefully. Some contain calcium chloride which will encourage the corrosion of reinforcement. Polymers may be added to these mixes to assist in bonding and to further close the pore structure.

Limes

The relevant British Standard is BS 890: 1972 *Building limes*; Part 1 *General*, Part 2 *Hydrated lime*, Part 3 *Quicklime*, Part 4 *Lime putty*.

Limes are being specified more as architects and builders realize the problems in specifying too strong a mix of cement, especially in the rehabilitation of existing buildings where over strong cements can cause cracking and subsequent water penetration. There has been resistance within the building industry to the use of limes which can cause health problems if used and mixed without proper protective clothing. Lime can cause dermatitis and skin burns to the eye, mouth and nose.

Hydrated lime

Hydrated lime has four classifications:

Hydrated high-calcium lime (white lime)
Hydrated high-calcium by-product lime
Hydrated semi-hydraulic lime (grey lime)
Hydrated magnesian lime

Hydrated lime, or hydrate of lime, sometimes known as

caustic lime, is basically calcium oxide made from calcium carbonate which has been calcined or roasted to remove water and carbon dioxide at a temperature of about 1000°C. The resulting clinker formed in this process is then ground down to a fine powder. When recombined with water there is a violent reaction and the mixture is brought to boiling point. The combination of heat and caustic material is extremely dangerous and the mixture has to be made in steel tanks by people who are well protected, with eyes and hands covered. The mixture is then normally left to mature for two weeks at least before use and is then combined with sand to make mortars. Case hardening of the exposed surface of mortar takes place as water evaporates and the calcium oxium combines with carbon dioxide in the atmosphere to reform calcium carbonate. The inner mortar stays softer.

Most lime is sold ready slaked as *hydrated lime*, but care is still needed in its use as the high alkalinity will cause major skin irritation and can still be harmful to the eyes.[3]

High alumina cement (HAC)

High alumina cement used to be popular to specify because of its high early strength and its resistance to chemical attack. However, it undergoes mineralogical changes, known as *conversion*. When conversion occurs the cement is then increasingly vulnerable to chemical attack and can be substantially weakened. There are cases where collapse of beams made with HAC has been catastrophic. These included the assembly hall roof of a school in Camden (built 1954–5, failed June 1973, after 18 years), the geography reading room in the University of Leicester Bennett Building (built 1963–4, collapsed June 1973) and the roof of the swimming pool hall at the Sir John Cass Foundation and Red Coat Church of England secondary school (collapsed 1974). In all these cases failure was catastrophic and without warning. The conversion of the beams involved was a major factor although there were also detailing problems with insufficient bearing for roof beams, coupled with unreinforced bearing nibs. Checks had been carried out in Stepney but rebound hammer test yielded false results of concrete strength of 52–83 N/mm^2 as the beams were *case hardened* even though the concrete internally had converted to a far lower strength of between 12.5 and 30.5 N/mm^2. These failures prompted an immediate change in the British Standard Code of Practice for concrete from major structural work to domestic use only, although it had been banned in Germany since 1962 and restricted in use in France since 1943.

Figure 3.7 A building site in Spetse, Greece. The remixing of lime and water has to be carried out on site in containers that can withstand the chemical reaction and the heat generated. Reuse of the old cast iron bath in Spetse is ideal for slaking lime.

The most complete report available is that from BRE, published by HMSO in 1984 entitled *High alumina cement concrete in existing building superstructures* (SO40) by S C C Bate. This gives immense detail on the materials, case studies of failures and recommendations for testing.

Conversion is chiefly a result of humidity and temperature and the initial water:cement ratio. Depending on the levels of these two factors the inevitability of conversion is then determined by time. Monocalcium aluminate decahydrate is the main reaction product formed in the initial set of HAC. It is not stable in buildings with high temperatures and gradually converts to tricalcium aluminate hexahydrate. If the rate of the initial reaction is not controlled, the high temperatures reached will start this conversion process. The new compound formed is dense, taking up a smaller volume. This makes the overall microstructure of the concrete unstable, as large voids are formed, increasing overall porosity and lowering strength. Conversion is accelerated by the presence of sodium or potassium hydroxide solutions which may be introduced through water leakage or from the aggregates. These alkali hydroxide solutions can dissolve and decompose hydration products in HAC forming alkali aluminate and calcium hydroxide. If carbon dioxide and water are available calcium hydroxide and alkali aluminate can then be changed to hydrated alumina and calcium carbonate. This carbonate can then react again with the hydrated cement compounds. This leads to progressive deterioration of the main calcium aluminate hydration products of HAC. This form of carbonation leads to disintegration of the concrete, whereas in Portland cement concrete strength will increase. Carbonation can be far more rapid and proceed at a rate of 4 mm per annum irrespective of the amount of conversion taking place.

HAC is more vulnerable to sulphate attack after conversion, producing the greater volume material ettringite, and consequent cracking. Steel is also more vulnerable to possible corrosion in high-alumina cement.

See BRE IP 22/81 *Assessment of chemical attack of high alumina cement concrete* which advises on testing procedures, and BRE IP 8/88 *Update on assessment of high alumina cement concrete*. This document proves that almost full conversion can occur within 20 years and the original banning of the material is fully justified.

Recent research by the BRE[4] shows that if HAC is mixed with pozzolanic or latently hydraulic materials (slag) genhlenite hydrate is produced instead of aluminium hydrate (which is the compound that converts and causes strength loss).

3.5 Testing of cement

There is a new European Standard, EN 196 *Methods of testing cement standard*. It requires all European member countries to comply, and conflicting British Standards to be withdrawn. The differences with regard to Britain are relatively minor but the standard BS 4550 *Concrete cube test and the mortar cube test* will be replaced by a mortar prism test bringing Britain into line with other EC countries. Specimens for the mortar prism test will be 400 mm × 40 mm in cross-section and 160 mm long. The mix design will be 1 part cement, 3 parts sand and 0.5 parts water. Vibrating tables will be used to compact the mortar.

Specimens will be cured in water and then broken in half under flexure. The resulting two halves will then be subjected to compression testing across the 40 mm width. There will be a transition period from April 1991 for six months for the changeover, during which time results will be accepted from both methods of testing. Extensive laboratory testing has proved a definite mathematical relationship with regard to differences in strength between the two testing methods as shown in Table 3.1.

The formula for the general relationship between the mortar prism and concrete cube compressive test results is:

$$\text{Log}_e\ (y/x) = 0.28/d + 0.25$$

x = concrete cube compressive strength N/mm^2
y = mortar prism compressive strength N/mm^2
d = age at test in days

This established relationship between the two testing methods allows the use of a graph to make conversions especially for converting concrete cube strength at three days to the European standard requirement for measuring early strength at two days.

Table 3.1 Mortar prism and concrete cube test results

Age at test (days)	y/x
2	1.48
3	1.41
7	1.34
28	1.30

See following text for values of x and y.

3.6 Durability of concrete and mechanisms of failure

Although concrete is a substantial material with potentially a life between 60 and 100 years, it can deteriorate very rapidly in certain conditions. Rainwater containing carbonic or sulphurous acids can dissolve it directly, or cause swelling in the case of sulphurous acids. It can also be subject to frost action and consequent spalling. On top of this the twin mechanisms of carbonation and chloride attack, which can accelerate the corrosion of reinforcement, can also dramatically shorten the life expectancy of the material. Chloride content can also make concrete more susceptible to sulphate attack.

A durable concrete is one with a permeability which will limit penetration by water, oxygen and carbon dioxide. Durability should be obtained by using a normal weight aggregate, sufficient cement content, a low free water:cement ratio and a grade above C20. This assumes adequate cover which will be determined by exposure and fire resistance; and proper hydration. For a short guide to the durablity of mixes related to exposure conditions that takes account of all these factors see Table 3.2. Refer also to Tables 3.4 and 3.5 of BS 8110 Part 1: 1985 for recommendations with regard to cover.

Alkali–aggregate reactions

However substantial concrete as a material may appear to be, its eventual failure is often due to a number of chemical reactions that may destroy its structural integrity. Its ability to take part in chemical reactions show how it cannot be regarded as an inert material and, to some extent, all aggregates are reactive. Apart from deterioration by contact with external phenomena even its very constituents can react unexpectedly with each other. The particles or aggregates can have reactions with each other or with the cement. Generally there can be reactions between siliceous constituents in the aggregates and the alkalis, sodium and potassium hydroxide released in the hydration of Portland cement.

Table 3.2 Durability of unreinforced concrete

Conditions of exposure	Concrete not containing embedded material		
	Maximum free water/ cement ratio	Minimum cement content* (kg/m^3)	Lowest grade of concrete
Mild	0.80	180	C20
Moderate	0.65	275	C30
Severe	0.60	300	C35
Very severe	0.55	325	C35†
Extreme	0.50	350	C45

Note: The lowest grades of concrete may be reduced by not more than 5 N/mm^2, provided there is evidence showing that with the materials to be used, this lower grade will ensure compliance with the required minimum cement content and maximum free water/cement ratio.
* Inclusive of g.g.b.s. or p.f.a. content.
† Applicable only to air-entrained concrete.

Source: Table 6.2 BS 8110 Part 1:1989

'A gelatinous product is formed which imbibes pore fluid and in so doing expands, inducing an internal stress within the concrete. The gel will cause damage to the concrete only when the following three conditions occur simultaneously'[5] — these conditions are:

- A reactive form of silica is present in the aggregate in significant quantities.
- Sodium, potassium and hydroxyl ions are present in the pore solution within the concrete.
- Water is available.

These kinds of reactions were first identified in the United States in 1940 and in a dam in Jersey in 1971. In the British Isles most structures with this kind of reaction appear to be confined to the South West and Midlands in England. Structures suffering from ASR (alkali–silica reaction) are less than 50 years old and most are less than 20 years old.

This reaction often starts in wet concrete and if it completes by the time the hardening process has finished the resulting expansion, which causes cracking, will have been accommodated.

These reactions can occur and not be serious in affecting the strength of the concrete if the gel formed from the reaction can flow into cracks or voids. If there is no take up of this extra material then the pressure generated from the volume change can cause cracking. Resultant map cracking is generated on the surface of the building, possibly exuding a jelly material which solidifies as hard beads on the surface of the material. Map cracking is generated in unrestrained concrete whereas it is characteristic for cracks to run parallel to the reinforcement in restrained concrete. The cracks are local to the aggregate reaction and usually confined to between 25 mm and 150 mm on the exposed concrete surface with cracks extending between 25 to 50 mm from the surface. The reaction is finite and will stop if reactive silicas or alkalis are depleted. Consequently cracking due to these reactions often shows within about $3\frac{1}{2}$ years.[6]

There are three main reactions that can occur:

Alkali–silica reaction
Cement–aggregate reaction
Alkali–carbonate reaction

The aggregates most likely to react are:

opal, chalcedony, microcrystalline to cryptocrystalline quartz, crystalline quartz (fractured or strained) latitic or adensite glass (cryptocrystalline devitrification products of those glasses)

The rocks that will contain these reactive minerals include:

cherts, siliceous limestones and dolomites, sandstones, quartzites, rhyolites, dacites, andesites, shales and phyllites[5]

To try and stop the reaction means careful choice in the selection and possibly the testing of aggregates. Porous aggregates could be chosen deliberately to house any likely reactive products. In addition, care must be taken in the specification of cement and controlling the amount of reactive alkalis available for any reaction, which should not exceed the 3.0 kg/m^3 limit. (See Concrete Society Technical Report 30, *Alkali–Silica reaction-minimising the risk of damage to concrete. Guidance Notes and model specification clauses.*) The alkaline character of the concrete depends on the concentration of potassium and sodium metals in Portland cement and the total cement content. As a guide, the concentration of these metals is usually expressed as an equivalent to sodium oxide (Na_2O). For example:

$$\% \text{ equivalent } Na_2O = \% Na_2O + \% K_2O \times 0.658$$

British Portland cements range from 0.4 to 1% equivalent sodium oxide (Na_2O) concentration. With equivalent Na_2O of less than 6%, cements are less likely to suffer from this reaction. Low alkaline cement can be specified by referring to BS 4027.[7]

De-icing salts can initiate alkali–carbonate reactions by bringing alkali ions into close proximity with the concrete. The rate of reaction can be very slow but it does affect engineering structures such as bridges; giving cause for concern because of the public safety dimension. These reactions have been a cause of major failure and are described in detail to emphasize that careful specification and purity of aggregates is essential.

Aggregates unlikely to be reactive are:

basalt, diorite, dolerite, gabbro, gneiss, granite, limestone, marble, micro granite, quartz, syenite. trachyte, tuff aircooled blast-furnace slag to BS 1047, expanded shale or slate, and sintered PFA[7]

There are laboratory tests that can be carried out to establish the potential reaction of cement–aggregate combinations. Sample bars are made up and examined for surface cracking, and any exudations of gel, before analysis of the interior by a petographer for any gel-filled pores.

Durability is affected chiefly by density of the mix and cover to reinforcement. Table 3.3 gives a guide to nominal cover from mild to extreme conditions with cover varying from 20 mm to 60 mm depending on the grade of concrete used. As durability is a function of the density of the material, note that this table refers to normal weight aggregates only.

Often the deterioration of concrete is complex and due to a number of combined causes. From a case study on the deterioration of concrete in a swimming pool in Sweden, apart from the basic alkali silica reaction detailed analysis showed other decay mechanisms had been initiated. Sulphates had formed producing ettringite and gypsum, pyrites had leached out and the feldspar content in the aggregates had disintegrated. As water had penetrated the concrete the deterioration had been so extreme that hydrated cement had leached out of the concrete and the material left was porous, with a low cement content and consequently low strength.[8]

See ASTMS Standard test methods:

C227 *Potential alkali reacting of cement–aggregate combinations*

C289 *Standard test method for potential reaction of aggregates* (chemical method)

The use of pulverized fuel ash (PFA) does seem to act as an inhibitor to this kind of reaction.[9]

Table 3.3 Durability cover and density of mix

Conditions of exposure	Nominal cover (mm)			
Mild	20	20*	20*	20*
Moderate	35	30	25	20
Severe	—	40	30	25
Very severe	—	50†	40†	30
Extreme	—	—	60†	50
Maximum free water/cement ratio	0.60	0.55	0.50	0.45
Minimum cement content (kg/m^3)	300	325	350	400
Lowest grade of concrete	C35	C40	C45	C50

Note: This table relates to normal-weight aggregate of 20 mm nominal maximum size.
* These covers may be reduced to 15 mm provided that the nominal maximum size of aggregate does not exceed 15 mm.
† Where concrete is subject to freezing whilst wet, air-entrainment should be used.

Source: Table 4.8 BS 8110 Part 1:1989

Sulphate attack

Sulphate attack is due to physical and chemical processes. Sulphate ions can react with hydrated calcium aluminates to form ettringite, and combine with free calcium hydroxide to form gypsum. The new compounds produced have a greater volume which leads to cracking and can be more easily dissolved leading to breakdown of the material. In order to minimize sulphate attack mixes should be controlled so that there is less free lime available for reaction, and the mix results in less permeable concrete. This can be achieved by using PFA which will combine with free lime as well as decreasing permeability. Alternatively OPC can be used if the tricalcium aluminate content is reduced to 3.5% by weight, which is the main compound which will react with sulphates.

Sulphates can be present in groundwaters due to the sulphate content in subsoils. They fall into three main categories:

Calcium sulphate (gypsum or selenite)
Magnesium sulphate (Epsom salt)
Sodium sulphate (Glauber's salt)

These compounds occur in particular stratas of London clay including bands of Lower Lias, Oxford clay, Kimmeridge clay and Keuper Marl. If building is carried out on previously developed land there may be contamination from works processes, particularly if these interact with pyrites in the soil (iron sulphide). There may also be contaminated fill from colliery shale below concrete floors. Brick rubble used as hardcore may have plaster debris attached which will be another source of calcium sulphate. Concrete used in conditions which have high sulphate content should be well compacted, highly durable and use sulphate-resisting cements. Admixtures which reduce water content, and so decrease permeability, are advantageous. Chemical analysis of ground conditions should take place before specifications are decided and in order to avoid cracking. Areas particularly prone to settlement there should be checked for sulphates in groundwater with a benchmark concentration of 500 parts per 100,000 in groundwater and 2.5% SO_3 in soil. If this is the case preventative measures should be taken by changing the nature of the cement and possibly using bitumen barriers in addition.

There are five classes of soil conditions which increase in value dependent on the sulphate content found in soil and groundwater.

See BRE Digest 250 *Concrete in sulphate-bearing soils and groundwaters*, revised 1986, for more detailed specification advice dependent on the class of sulphate concentration found.

Carbonation

Carbonation is a term given to concrete that is affected by the diffusion of carbon dioxide through the cement gradually converting calcium hydroxide to calcium carbonate. The rate of diffusion is affected by the permeability of concrete and the ultimate effect is to reduce the pH of the concrete from a protective alkaline level of 13.4 to an acidic environment of as low as 3.6, which is then more conducive to the corrosion of reinforcement. Steel is less likely to corrode in a passive alkaline environment which inhibits the passage of free ions. As concrete has deteriorated, investigations have shown that this is one of the most damaging mechanisms affecting the long term durability of concrete. As knowledge is gained about this phenomena the standards for increasing concrete quality and depth of cover have changed in response. Carbonation can be detected by an on site chemical test using phenolphthalein as a detector. If it turns bright pink on application to a concrete sample it shows the concrete has still retained some protective alkalinity. Concrete that has suffered carbonation shows no change of colour. (See BRE Information Sheet 14/78 on the technique for on-site testing.)

Concrete with low water:cement ratios does not show the kind of pore structure that facilitates the easy diffusion of carbon dioxide. There is also dependency on the relative humidity of the atmosphere which encourages carbonation if between 50 and 65%. Diffusion is more rapid in large dry capillary pores in concrete, and is inhibited in capillary pores that are fully saturated with liquid. Other factors include temperature dependency and whether there are local cracks in the concrete which will provide a faster route for carbonation where conversion of material will take place at right angles to fractures.

Chloride attack

The direct result of chloride attack is the corrosion of reinforcement in concrete. If reinforcement has corroded it will be detectable by the appearance of cracks running parallel to the main reinforcement bars or to stirrup bars. This will probably lead to the spalling of concrete.

If there are free chloride ions available, although a passive oxide layer may have formed on the outside of reinforcement, the very presence of chloride ions within this film will increase the electrical conductivity of the film, sometimes even preventing the formation of this passive layer but certainly enabling the corrosion process to continue. Pitting corrosion is characteristic of chloride attack, as ferrous ions are formed which migrate from the site of the original attack but still form adjacent corroded deposits. This leaves the original source of corrosion activity *live* to produce greater numbers of ferrous ions by dissolution of the steel. This type of corrosion is also dependent on the pH level of the concrete and there are Pourbaix diagrams available which show the likelihood of corrosion dependent on the twin variables of electrode potential and pH of the immediate gel pore environment. If the pH level falls due to carbonation then corrosion is more likely to take place, which is why carbonation and chloride attack are usually joint mechanisms that have led to the overall deterioration of existing concrete structures.

Chlorides come from a number of sources. They can be present in marine aggregates which have not been properly washed, or transported via a saturation of the concrete in sea water conditions or where roads have been salted to prevent ice formation. An alternative to the use of sodium chloride is now calcium magnesium sulphate (CMA), an organic bio-degradable non-corrosive icing salt, now being used extensively in the US Federal Highway Programme. This compound is used as a scrubber for coal-fired power stations, producing calcium sulphate in the process. It was originally developed as an alternative to sodium chloride to avoid contamination of local water supplies and is processed from renewable woody biomass. A large source of contamination is from the incorporation of calcium chlorides in concrete mixes when used as an accelerator. The use of these accelerators is now banned but the repercussions of their use has been the large-scale corrosion of reinforcement in concrete structures which now require extensive repair.

Preventative measures for corrosion would include the specification of a durable concrete mix with adequate cover and possibly the use of reinforcement which has the additional protection of hot dip galvanizing or an epoxy coating. In very exposed situations, such as marine atmospheres or saturated conditions (as in oil rigs), it may be worth using stainless steel reinforcement. Stainless steels can still be vulnerable to corrosion and those specified should fall into the category of stabilized steels which precipitate molybdenum carbide instead of chromium carbide.

Repairs can be carried out which include the cleaning of corroded reinforcement and sometimes the welding in of new pieces. The steel is then coated with an epoxy resin and the cover to the reinforcement made good. It may be advisable to coat concrete with a paint finish that will minimize the diffusion of carbon dioxide and so add to the longevity of the concrete. (See BRE Information paper 12/80, *Deterioration due to corrosion in reinforced concrete*.)

There is an on site portable test developed by the Concrete and Materials Laboratory of British Rail for measuring the chloride content. A powdered 5 g sample is obtained by drilling and combined with 50 ml nitric acid. This releases all the chloride content which can then be evaluated. Results obtained from this test can then be correlated with the depth of carbonation to establish the likelihood of reinforcement being at risk from corrosion, which in turn will give guidance as to the remedial measures needed.

Frost

The attack of concrete by frost is well documented with the formation of ice lenses and also the initiating of cracking due to freezing conditions. Cracking may be due to the hydraulic effects of pore water or by the expansion of ice. This is a major problem and has to be taken into into account in casting concrete in poor weather conditions where steps must be taken to ensure controlled curing conditions. The pore size of concrete is also critical with regard to these effects (see section 3.10, page 44, on admixtures and the control of pore size). Recent research is starting to look at concrete not just as a material that has a known strength that may then be affected by freezing conditions, but as a composite. This composite has a concrete matrix with air, water or ice phases that have a calculable and predictable elastic modulus, for example, using volume fractions in the law of mixtures. In this scenario it has been proved that water and ice together have load bearing properties and the effects on strength can be predicted.[10]

3.7 Mechanical behaviour of concrete as a composite

For a composite to be successful there must be good adhesion between the fibre or particle and the matrix. Concrete relies partly on good mechanical bonding achieved in the kind of surface irregularities found on reinforcement, as well as some weak chemical (hydrogen) bonding between cement and the aggregate interface. Cracking is more likely to propagate at these interfaces and this can be minimized by choice of aggregate. If aggregates are chosen that are angular from crushing rather than smooth and rounded from the wearing action in running river beds, the greater mechanical degree of interlocking will improve the overall strength of the concrete. There can be plastic cracking of concrete in the initial set from shrinkage of the hardening material or from settlement of aggregates and cement and loss of water (often referred to as bleeding) in deep sections. Thermal cracking may arise from great temperature differentials through the hardening slab and the outside air temperature. The only acceptable level of cracking is a dimensional gap of 0.3 mm or below. Wider gaps will significantly affect water penetration and the corrosion of reinforcement and durability generally.

Fracture toughness is also affected by aggregate size. The strength of concrete can be shown to increase as the size of the aggregate increases. There are two main reasons for this. If a crack grows it tends to propagate around aggregate particles and more energy is needed for an increased path length of fracture. As stress is applied there is enough displacement of particles so that frictional forces become important, there is greater cohesion and resistance to the force with the result that fracture may then occur at a higher stress as the material behaves more plastically. Alternatively as the crack finds difficulty in propagating through the matrix it then passes through the actual aggregate, again at a higher stress.[11] The fracture energy needed in concrete made with aggregate sizes of 12–16 mm increased seven fold from concrete made with aggregate sized 2–4 mm. As a general observation, concrete should be made with the maximum sizes of aggregate available.

In practice concrete tends to fracture at lower values than its theoretical strength. This is due to a number of micro-defects such as scratches between aggregate and mortar or small holes that act as stress concentrators.

Monitoring of concrete strength if critical, is now made easier by the invention of *smart concrete* which incorporates 2% by volume of carbon fibres. These increase the conductivity of concrete by a factor of 10 and any changes in stress patterns can be detected as the concrete's electrical resistance changes.

Fire resistance

Concrete is used as cover to fire-protect steel and is effective because of its density and inert behaviour as a ceramic material. In looking at BS 8100 for guidance as to the cover required up to 4 hours fire resistance will be achieved with a minimum cover of 80 mm; see Table 3.4. The cover required varies with regard to the structural element being protected and beams may need a greater depth of cover than floors for fire resistance over $1\frac{1}{2}$ hours.

Under fire load concrete may expand and this should be catered for in the detailed structural design of a building. Section 4 of BS 8100 gives a complete guide to fire resistance and the additional protection obtained by other

Table 3.4 Fire resistance

Fire resistance	Nominal cover (mm)			
	Beams		Floors	
	Simply supported	Continuous	Simply supported	Continuous
0.5	20	20	20	20
1.0	20	20	25	20
1.5	35	20	30	25
2.0	60	35	40	35
3.0	70	60	55	45
4.0	80	70	65	55

Source: Table 4.9 BS 8110 Part 1:1989

finishes. Tables are given which show the dimensions of elements related to the fire resistance needed and the depth of cover required.

Elastic modulus

The elastic modulus of concrete is mostly determined by the aggregates used. The figure for an individual mix is derived from the characteristic cube strength at 28 days modified by two constants (see section 7.2 of BS 8110).

$E_{c,28} = K_0 + 0.2F_{cu,28}$
$E_{c,28}$ = static modulus of elasticity at 28 days
$F_{cu,28}$ = characteristic cube strength at 28 days
K_0 = constant related to the modulus of elasticity of the aggregate which is taken at 20 N/mm^2 for normal weight concrete.

Values can be found for aged samples which will have increases in cube strength over time.

As concrete is a composite, the homogeneous strain model developed by Voigt provides a simple formula for establishing a composite modulus using the law (or rule) of mixtures.

$$E_c = E_1V_1 + E_2V_2$$

where V_1 would be the concrete volume and V_2 the steel volume.

Thermal movement

Concrete expands on being heated and also when made wet; conversely it contracts on cooling and when it loses water. Movement joints have to be built in to accommodate these thermal changes which can be in the order of 10 mm over a 30 m length of member. If movement of the concrete was restrained there would be tensile forces which would cause cracking. The effect of reinforcement is to distribute the cracking pattern and reduce the effect of the internal stresses.

Cracking can occur in concrete from high temperatures due initially to the heat of hydration and then a fall back to the ambient temperature causing initial contraction and flexural stress. Seasonal variations in temperature can widen these initial thermal cracks. Any exposure to earth or contact with ground in temperate climates means that the concrete will never dry out and the range of movement with changes in temperature in these conditions will be greater.[12]

Contraction can be minimized by:

- Using aggregates with low-medium coefficients of thermal expansion/shrinkage. (This indicates the use of seasoned crushed rock or river/marine aggregates that are environmentally stable.)
- Minimizing the cement content which should not exceed 400 kg/m^3.
- Using concretes which do not exceed grade.
- The controlled curing of concrete and prevention of water loss.
- Temperature control in curing.

Shrinkage

This varies according to section size and relative humidity (RH). Shrinkage increases as the humidity lowers, reaching a peak at about 20% RH. There will be cyclical shrinkage every year working out at approximately plus or minus 0.4 of a 30-year shrinkage value, given in Figure 7.2 of BS 8100 Part 2: 1985. Typically this would give for a 300 mm section at 30% RH a shrinkage after 30 years of 3.75×10^{-4} m (0.000375 mm). Shrinkage in the initial set of concrete is more dramatic and can give a dimensional change, for example, of 1.5 mm over a 3 m length of unreinforced concrete.

3.8 Theory of mixes

Taking the previous principles into account, the design of a suitable concrete mix relies on the results of experience and tested strengths. BRE Digests 325 and 326 deal with the materials used for making concrete and their specification, design and quality control. BS 5328: 1981 *Methods for specifying concrete, including ready mixed concrete* is the basic standard for referral although it is limited in its guidance for mix design, which confines concrete produced to grades below C30. As the variables are so great, concrete mixes have a designed strength known as the characteristic strength which is much greater than the required strength, which is known as the target strength.

All mixes are referred to by grades:

C7.5, C10, C15, C20, C25, C30, C35, C40, C45, C50, C55, C60, F3, F4, IT2, IT2.5, IT3

They relate to the following strengths:

C = characteristic compressive strength (N/mm^2)
F = flexural strength (N/mm^2)
IT = indirect tensile strength (N/mm^2)

A 'P' after the grade, e.g. C20P, indicates a prescribed mix.

Concrete between grades 7 and 10 is referred to as plain concrete. Grade 15 will be reinforced but use lightweight aggregate. Grades 20–25 will be reinforced with dense aggregate. Grade 30 will use post-tensioned reinforcement and grades 40–60 will use pretensioned reinforcement. Table 3.5 summarizes the main grades and their strengths.

Mixes will fall into the following categories:

Designed
Standard
Prescribed

Standard mixes will be selected from BS 5328 and will not include any admixtures. They are for small jobs and are limited to the lower grades.

Strength

The indirect tensile strength of concrete, sometimes called the flexural strength, is approximately one tenth of the compressive strength. This ratio is variable according to the strength of the aggregates specified.

The modulus of elasticity of concrete is dependent on the composite nature of the material. Moduli for composites are normally worked out on the basis of the fractions of different materials. Hence there is not a constant modulus for concrete. It varies according to the crushing strength of the concrete specified and will alter as the strength of aggregates and other aspects of the mix varies, i.e. from 25–36 kN/mm^2. Some typical ranges of the static modulus of elasticity are given in Table 3.6.

Table 3.6 Static modulus of concrete

$f_{cu,28}$ (N/mm^2)	$E_{c,28}$ (kN/mm^2)	
	Mean value	Typical range
20	24	18–30
25	25	19–31
30	26	20–32
40	28	22–34
50	30	24–36
60	32	26–38

Source: Table 7.2 BS 8100 Part 2:1985

Prescribed mixes and designed mixes

There are two ways of specifying a concrete mix. The first is to use ordinary *prescribed mixes* which already have the proportions of cement and aggregates worked out relative to a required strength, i.e. the grading is specified (see Table 3.7). There can be special prescribed mixes where an aspect or aspects of the mix control will change rather than the constituents, i.e. workability, maximum free water:cement ratio, air content, temperatures and admixtures. Most building professionals will use prescribed mixes which have already been worked out and the proportion of constituents already fixed. They can be specified by *grade* which relates numerically to the characteristic strength.

The second is to use a *designed mix*, which is a performance specification stating required strength and minimum cement content but leaving the grading and details of the mix design to be worked out.

The BRE report *Design of normal concrete mixes*, 1988, takes users through a process which works in clear stages (see Table 3.8):

(1) Strength and water:cement ratio.
(2) Workability and free water content.
(3) Stages 1 and 2 combine to give cement content.
(4) Total aggregate content.
(5) Selection of coarse and fine aggregates.

Table 3.5 Strength of concrete

Grade	Characteristic strength f_{cu} (N/mm^2)	Cube strength at given age (N/mm^2)				
		7 days	2 months	3 months	6 months	1 year
20	20.0	13.5	22	23	24	25
25	25.0	16.5	27.5	29	30	31
30	30.0	20	33	35	36	37
40	40.0	28	44	45.5	47.5	50
50	50.0	36	54	55.5	57.5	60

Source: Table 7.1 BS 8100 Part 2:1985

Table 3.7 Prescribed mixes

Concrete	Constituents		Mix A	Mix B
7	Cement (kg)		210	230
	Total aggregate (kg)		1900	1800
	Fine aggregate (%)		35–50	35–50
10	Cement (kg)		240	260
	Total aggregate (kg)		1850	1800
	Fine aggregate (%)		35–50	35–50
15	Cement (kg)		280	310
	Total aggregate (kg)		1800	1750
	Fine aggregate (%)		35–50	35–50
20	Cement (kg)		320	350
	Total aggregate (kg)		1800	1750
	Sand — grading limits	C	40	45
		M	35	40
		F	30	35
25	Cement (kg)		360	390
	Total aggregate (kg)		1750	1700
	Sand — grading limits	C	40	45
		M	35	40
		F	30	35
30	Cement (kg)		400	430
	Total aggregate (kg)		1700	1650
	Sand — grading limits	C	40	45
		M	35	40
		F	30	35

Mix A: medium workability; 25–75 mm expected limit to slump.
Mix B: high workability; 75–125 mm expected limit to slump.

Source: Table 9 *Concrete practice* 1987 Cement and Concrete Association

If a mix is designed from the basic ingredients, trial mixes need to be carried out and tested.

Engineers can specify mixes outside this range for special situations. All ingredients are now weighed in line with current American and European standards. Volume batching is not so accurate and not recommended as, although minimum strengths may be obtained, the resulting thinner mix may not be as dense nor as durable.

Grading

For strength, as a composite, concrete must have a closely controlled packing arrangement. Using the analogy of six tennis balls in a box, the interstices must be filled with smaller particles, and the new interstices created filled with even smaller particles. The whole packing arrangement must allow for every piece of aggregate to be surrounded by the matrix of cement. Concrete is often seen in drawings in elevation or section, and the three-dimensional nature of linking each particle to its neighbour by *wrapping* is lost.

Packing arrangements are specified by the *grading* of aggregates. Aggregates are divided into two broad categories, fine and coarse. These are defined by sieve sizes to BS 410. The maximum size of the coarse aggregate should not be more than a quarter of the minimum thickness of the concrete section. The BS sieve sizes are:

Coarse 75, 63, 37.5, 20, 14, 10, 5 mm
Fine 2.36, 1.18 mm and 600, 300, 150 microns

Preferred sizes usually fall into the following range of sieve sizes: 40, 20, 14, 10.

For a mix to be continuously graded means there is a complete range of sizes in the mix. Sometimes there will be *gap grading* which is not advantageous, leading to voids. It is unusual to use aggregates over 40 mm for mass concrete, and to use aggregates over 20 mm for reinforced work. Overall control of aggregates is needed from one source of supply to obtain consistent colours. Aggregates are sieved and grading is calculated as the percentage by weight of material passing through the different sized sieves.

Workability

The workability of a mix is controlled not just by the water content but also by the specification of aggregates. Finer particles move in a more fluid manner and increase the workability and placing of a mix but require more water which will affect the resulting strength. The cement content in a mix with high workability is always higher. Particle configuration can also affect workability. River dredged aggregate, which is smooth, is more workable than angular crushed rock particles. Air entrainment can also be introduced to increase workability which may allow a reduction in fine aggregate, up to 5% of the total aggregate used, saving on some of the water content. The slump test is most widely used to check workability.

Aggregates

Apart from grading in size, aggregates can be categorized by their geological origin and by their density. As concrete mixes become more precise, and demands are made on the quality of concrete for greater durability, this reflects on the sorting, classification and quality of aggregates used which make up the bulk of the mix. Some aggregates, notably dolerite and whinstone, can have a higher initial drying shrinkage, and this can alter the calculated water content of the mix.

As defined by BS 3797, which uses bulk density as a measure, lightweight aggregates have a density of up to 1200 kg/m^3 for fine grades, and 960 kg/m^3 for coarse grades. Up to 4000 kg/m^3 they are termed *normal density*

Table 3.8 Example of concrete mix design (20 mm nominal maximum size of aggregate

Stage	Item*		Reference or calculation	Values
1	1.1	Characteristic strength	Specified	25 N/mm^2 at 28 days Proportion defective / %
	1.2	Standard deviation	Fig 3	/ N/mm^2 or no data / N/mm^2
	1.3	Margin	C1 or Specified	(k = /) / × / = / N/mm^2 10 N/mm^2
	1.4	Target mean strength	C2	25 + 10 = 35 N/mm^2
	1.5	Cement type	Specified	OPC/~~SRPC~~/~~RHPC~~
	1.6	Aggregate type: coarse Aggregate type: fine		~~Crushed~~/uncrushed ~~Crushed~~/uncrushed
	1.7	Free-water/cement ratio	Table 2, Fig. 4	0.57 } Use the lower value: 0.50
	1.8	*Maximum free-water/cement ratio*	*Specified*	0.50
2	2.1	Slump or Vebe time	Specified	Slump 30 - 60 mm or Vebe time / s
	2.2	Maximum aggregate size	Specified	40 mm
	2.3	Free-water content	Table 3	160 kg/m^3
3	3.1	Cement content	C3	160 ÷ 0.50 = 320 kg/m^3
	3.2	*Maximum cement content*	*Specified*	/ kg/m^3
	3.3	*Minimum cement content*	*Specified*	290 kg/m^3 use 3.1 if ≤ 3.2 use 3.3 if > 3.1 : 320 kg/m^3
	3.4	Modified free-water/cement ratio		/
4	4.1	Relative density of aggregate (SSD)		2.5 known/~~assumed~~
	4.2	Concrete density	Fig 5	2326 kg/m^3
	4.3	Total aggregate content	C4	2325 − 320 − 160 = 1845 kg/m^3
5	5.1	Grading of fine aggregate	Percentage passing 600 μm sieve	90 %
	5.2	Proportion of fine aggregate	Fig 6	20 to 23, say 22 %
	5.3	Fine aggregate content	C5	1846 × 0.22 = 405 kg/m^3
	5.4	Coarse aggregate content	C5	1845 − 405 = 1440 kg/m^3

Quantities	Cement (kg)	Water (kg or *L*)	Fine aggregate (kg)	Coarse aggregate (kg) 10 mm	20 mm	40 mm
per m^3 (to nearest 5 kg)	320	160	405	260	395	785
per trial mix of 0.08 m^3	25.6	12.8	32.4	20.8	31.6	62.8

* Items in *italics* are optional limiting values that may be specified.

Source: Table 5 *Design of normal concrete mixes* 1988 BRE

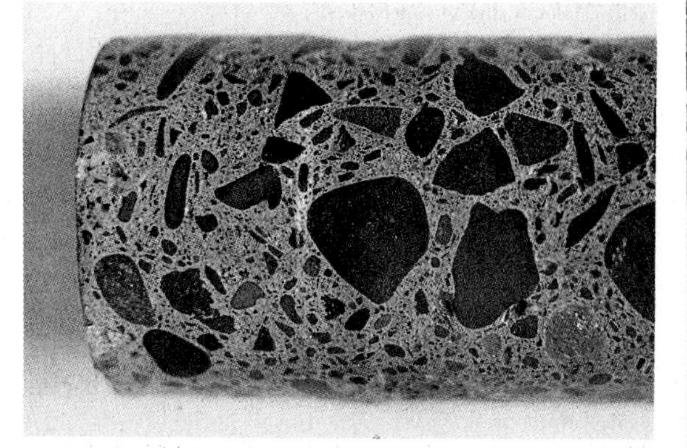

Figure 3.8 Core sample of concrete showing the variation in particle size.

aggregates, and over that value are referred to as heavy. Densities of 1200–2000 kg/m^3 are more likely, as this is the range for most rocks used.

Under BS 8110 aggregates are classified in terms of their particle density which raises the value of their grading. Lightweight aggregates are classified as those below 2000 kg/m^3, normal as between 2000 and 3000 kg/m^3 and heavyweight over 3000 kg/m^3. Most aggregates used will fall into the heavyweight category.

Most natural aggregates are quarried to yield gravels and sands without further processing, and marine and river bed aggregates only need grading. Igneous and sedimentary rocks are crushed to provide the right sizes of aggregates needed. These natural aggregates fall within the range of normal aggregates. Most aggregates used in the sedimentary category have been limestone aggregates. As the supply situation becomes more acute there is pressure to use lower grade material and sandstone aggregates. There are great sources of sandstone aggregate in the North West of England which have been evaluated by the BRE (See Information Paper 16, 1989.) and are likely to be an important future source of building material. These aggregates are generally more porous and will be lower strength. Their silicaceous nature does not appear to have significant reactions with cement.

Lightweight aggregates are consequently less dense than natural aggregates and are more likely to be from industrial waste products. The most used is pulverized fuel ash (PFA) from the burning of powdered coal in power stations. Natural lightweight aggregates include perlite, vermiculite, clays and shales which are heated to a point where their expansive open texture gives higher thermal resistance and great insulating values to concrete. These still have a sufficiently rigid structure to satisfy minimum crushing strengths.

There are a number of British Standards which cover the range of aggregates used.

Figure 3.9 Sea shore pebbles classified in concrete making as marine aggregates. Their characteristic round shape comes from being subjected to great turbulence and the wearing action from other stones.

- BS 877 *Foamed or blastfurnace slag lightweight aggregates for concrete.* (See section 5 of BS 8710 Part 2: 1985.)
- BS 882 *Specification for aggregates from natural sources for concrete.*
- BS 1047 *Aircooled blastfurnace slag. Coarse aggregate for concrete.*
- BS 3797 *Specification for lightweight aggregates for concrete.* (See section 5 of BS 8110 Part 2: 1985.)

Aggregates from recycling

There is a valuable resource in the recycling of rubble from buildings. It is now becoming more viable in Northern Europe, particularly in Germany and the Netherlands, in places where the recycling of rubble was used to rebuild the towns and cities damaged during the Second World War. The economics of waste disposal now presents a problem, and the initial expense of waste disposal, often by haulage to sites well out of cities as landfill sites are lost, is coupled with the long haulage of new aggregates, and the time and energy expended in transport. The total cost of this process can now be compared with reprocessing within the city, allowing for the collection, sorting and redistribution of building rubble. Critical distances which are starting to make recycling viable are of the order of 200 km in Northern Europe. The whole process needs awareness and the co-operation of local authorities. There is an investment in processing plant to consider which can use the same kind of crushing and screening facilities as are used for initial materials processing. This equipment is readily available (Klockner Becorit). In concrete the feed size can be as much as 1200 mm for reduction to below 200 mm. This will still yield 30% of particles of 8–52 mm.

Recycling is at present in the hands of private enterprise in Germany but it makes sense for a policy whereby local authorities could act as major centres for recycling and graded material would then be available for reuse. Tests in Germany for establishing standards for road construction material confirm the quality of the recycled rubble in respect of impact and aging resistance, Proctor density, moisture content, water permeability and California Bearing Ratio stability. It would also pre-empt criticisms from local authorities on the use of this material as they would be in charge of its overall quality. This has been a problem in Berlin.

As these waste materials (such as pulverized fuel ash from steel production as lightweight aggregate for concrete) become more acceptable, and the volume of recycled material grows, the pressure on the use of primary resources for aggregate material, which directly affect our own landscape, could be relieved.

Most recycled rubble is used for roads and foundations but, given a system of quality control and approved testing procedures, the aggregates should be available for wider use and possibly for normal structural concrete. The quality of the material generally should be higher than the primary aggregates, given the increased strength in the cement (doubling over 20 years). Although the ultimate strength could still be limited by the original aggregate, as concrete could be viewed as a three-dimensional cellular structure, it would be logical to experiment with mixes using large aggregates.

Waterproof concrete

Watertight Concrete Construction by R Colin Deacon (C & CA, 1984) sets out the basic principles for the specification and design of structures which need to be water retaining. BS 5337 referred to in this guide is now withdrawn and reference should now be made to BS 8110 to establish the exposure conditions for concrete. See Table 3.9.

Watertight is defined as 'stopping the flow of water although there may be some damp patches'. If these are unacceptable, additional tanking of the structure may be needed. General practice requires for a continuously graded mix aggregates which do not have a water absorption greater than 4%, generally avoiding limestone aggregate (see p. 48 for definition of watertightness and discussion of the control of water:cement ratio and admixtures). If limestone aggregate has to be used then it may need testing

Table 3.9 Exposure conditions for concrete

Environment	Exposure conditions
Mild	Concrete surfaces protected against weather or aggressive conditions
Moderate	Concrete surfaces sheltered from severe rain or freezing whilst wet Concrete subject to condensation Concrete surface continuously under water Concrete in contact with non-aggressive soil
Severe	Concrete surface exposed to severe rain, alternate wetting and drying or occasional freezing or severe condensation
Very severe	Concrete surfaces exposed to sea water spray, de-icing salts (directly or indirectly), corrosive fumes or severe freezing conditions whilst wet
Extreme	Concrete surfaces exposed to abrasive action, e.g. sea water carrying solids or flowing water with pH$\leq$4.5 or machinery or vehicles

Source: Table 3.2 BS 8110 Part1:1985

with regard to its strength and water solubility, especially in soft water areas. Greater control of cracking is needed and therefore a higher amount of reinforcement in the structure. The conditions for different mix design can also vary throughout the structure. Attention also has to be given to joint design which should be simple straight-through butt joints with water bars. Water bars are generally rubber or PVC and the PVC type have greater potential for long-term durability. Complex profiles should be avoided and the dumb-bell type is a satisfactory profile for most situations.

The specification of watertight mixes is defined by three classes of exposure: A, B and C. The highest class, A, will need a grade 30 mix which is adequate for high quality and for concrete reservoir roofs, with class B being adequate for buried walls in reservoirs using a grade 25 mix.

Sometimes the mix design should be looked at carefully to provide a different level of water resistance dependent on location. The area where walls hit the ground can be vulnerable to groundwaters and subsequent efflorescence.

3.9 Concrete quality and durability in general

To improve the overall quality of concrete, and also its durability and strength, the number of defects in a mix can be reduced in the following four ways:

- Milling the cement to a finer powder.
- Using the ideal water:cement ratio 0.38.

Figure 3.10 This ground detail of the Times building by Tadao Ando, which is next to a canal in Kyoto, shows the vulnerability of detailing walls at this low level which have to cope with external abuse as well.

- Adding polymeric lubricants (which allow the particles to pack more densely).
- Applying pressure during hardening (which squeezes out residual porosity).

Using these basic guidelines, results show that an overall porosity can be achieved of less than 2% with a tensile strength of up to 90 MPa. These improvements can be demonstrated by making a controlled mix and squeezing the mixture through an ordinary mangle to remove any air, and producing spring-like forms that are sufficiently elastic to be comparable to steel.

Reinforcement

Reinforcement is often thought of as a component in concrete that needs calculation for the right sizing and the correct cover. It also needs to be regulated by standards in terms of quality and fixing. However, designers lose sight of the fact that it is often a three-dimensional structure that needs continuity. This continuity is achieved by lapping and tying which involves bending the reinforcement and making junctions which take up space. Beams may be reinforced with *cages* where main bars are positioned within rectangular links or stirrups. The links are tied with mild steel wire which has to be kept clear of the formwork as it might penetrate through the concrete and rust. Section sizes may have to be altered to accommodate the bending of steel, as although the inside radius for bending is normally twice the bar diameter for mild steel, up to a section size of 20 mm in diameter, this can increase to four times the bar diameter for section sizes over 25 mm in diameter. Table 3.10 gives radii relative to bar diameters. This has to be allowed for to ensure that adequate concrete cover can still be maintained, and is usually carried out by a skilled steel fixer. The consulting engineer normally makes up detailed bar bending schedules which catalogue all the shapes required and their position. Bar bending should not be carried out if the temperature of the steel is below 5°C. It is permitted to heat the steel as long as this does not exceed 100°C.

Reinforcement varies as a result of its processing and can be *hot rolled, cold drawn* and either *mild steel* or *high tensile steel*. The processing and type of steel will give a range of strengths. The main British Standards are:

- BS 4449: 1988 *Carbon steel bars for the reinforcement of concrete*. This covers hot rolled and cold worked deformed bars.
- BS 4466: 1989 *Specification for the scheduling, dimensioning, bending and cutting of steel reinforcement for concrete*. See table 1 for a schedule of preferred shapes.
- BS 4482: 1985 *Specification for cold reduced wire for the reinforcement of concrete*. The sizes are 5, 6, 7, 8, 9, 10 and 12 mm (not for prestressing).
- BS 5896: 1980 *Specification for high tensile steel wire and strand for the prestressing of concrete*. The strands are made of 7 mm wire configurations with tensile strengths that vary between 1670 and 1860 N/mm^2.
- BS 4483: 1983 *Specification for steel fabric for the reinforcement of concrete*. The wire diameter varies between 5 and 10 mm diameter in increments of 1 mm and is cross-welded, with pitches of normally 100 or 200 mm. The sheet sizes come in standard lengths of 4.8 m and widths of 2.4 m.
- BS 8110 Part 1: 1985 Section 7 *Specification and workmanship of reinforcement*.

Welding reinforcement on site is not encouraged, only particular weldable steels can be used, and welding should be restricted to fixing reinforcement in position between crossing members, or transferring loads to other steel members.

The bonding of reinforcement with concrete is important and this can only be achieved if the surface of the steel is relatively clean and free from lost mill scale or loose rust. All bars will have some rust and this does not adversely affect their performance. Once in the passive alkaline environment of the concrete they should not decay further. If corrosion is an issue then austenitic stainless steel would be the most effective although expensive specification followed by epoxy coated steel and then galvanized steel.

Table 3.10 Radii relative to bar diameters

Bar size	Type and grade R and type and grade S			Type and grade T and type and grade S			Fabric complying with BS 4483			
d	*r*	*n*	*h*	*r*	*n*	*h*	*d*	*r*	*n*	*h*
6*	12	100	100	18	100	100	5	15	100	100
8	16	100	100	24	100	100	6	18	100	100
10	20	100	100	30	100	110	7	21	100	100
12	24	100	110	36	100	140	8	24	100	100
16	32	100	150	48	100	180	9	27	120	135
20	40	100	180	60	110	220	10	30	120	135
25	50	130	230	100	180	350	12	36	130	145
32	64	160	290	128	230	450	—	—	—	—
40	80	200	360	160	280	560	—	—	—	—
50*	100	250	450	200	350	700	—	—	—	—

Bend

Hook

Note: The use of hooks as end anchorages is not normally necessary with deformed bars.
* Non-preferred sizes.

Source: Table 3 BS 4466 Part 1:1989

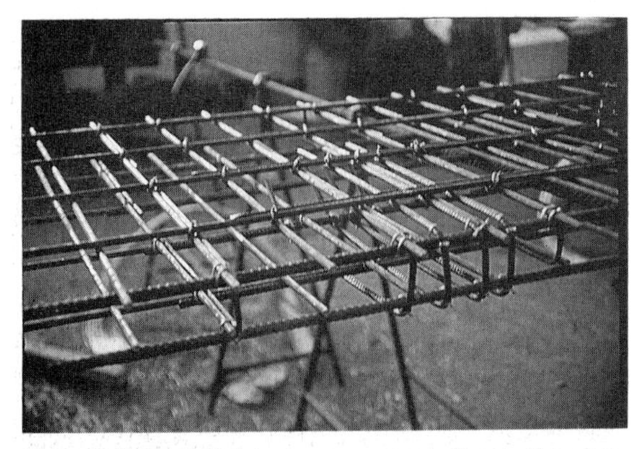

Figure 3.11 Reinforcement cage at Trent Concrete, Leicestershire. Although concrete is a cast material, as reinforced concrete it cannot be thought of in isolation from the three-dimensional structure of the steel. Here, the reinforcing cage is shown for a precast concrete panel.

The surface also should not have been contaminated with grease or mud, retarders, oils or other materials which could affect the bond strength. The correct storage of reinforcement is important and it should be stacked neatly off the ground. If delivered to site, allowance should be made for the fact that it is normally delivered in 12 m lengths.

Standard bar diameters These are of the following sizes (in mm):

Grade 250 (0.42% carbon) 8, 10, 12, 16
Grade 460 (0.51% carbon) 8, 10, 12, 16, 20, 25, 32, 40

Both grades also contain sulphur, phosphorous and nitrogen and the grades reflect the characteristic strength in N/mm^2.

Standard mesh sizes These come in the following sizes (in mm):

200 × 200 slabs
100 × 200 slabs and walls
100 × 400 roads, paving, ground floor slabs
100 × 100 sprayed concrete work, encased steelwork

Concrete cover provides several functions. An adequate depth of cover must be allowed so that concrete cannot spall off the edges; it must give adequate fire protection and protect reinforcement against corrosion. Cover has steadily increased as the effects of carbonation have become known, and the durability of concrete has had to improve with respect to chloride attack. Another factor may be surface treatment which could reduce the cover considerably. The depth of cover varies according to the exposure conditions, the free water:cement ratio, cement content and overall grade. It is often forgotten in detailing that links will wrap around the outside of the main reinforcement, and a common complaint is that there is insufficient room to position the reinforcement and provide adequate cover over the links. The nominal cover will vary from 25 mm in mild conditions (usually protected indoor environments) to 60 mm for extreme conditions where concrete is expected to be water or fluid retaining. See Tables 3.3 and 3.4.

There may be special needs which require the reinforcement to remain stable in critical conditions, such as in nuclear power stations or in ferro cement mixtures where mesh reinforcement will be close to the surface. In these cases stainless steel or coated carbon steel reinforcement may need to be specified. The costs are usually prohibitive for normal building practice.

Prestressed and post tensioned concrete

Both of these methods use the same principle, which is to put the reinforcement (which is usually in the form of wires or tendons) under tensile stress before it is finally encased in a matrix. The matrix is cast around the wires and the stress then released on the wires. As the wires contract they exert a sideways compressive force on the matrix, pre-stressing the material and enabling it to take a greater load. The prestressing allows a greater resistance to subsequent tensile stress as there will be an inherently greater internal resistance to force applied. It is a good demonstration of putting energy into a system which is then *frozen*, but effective in resisting further stress. The disadvantage of using prestressing systems arises when buildings using these systems are demolished. If pretensioned wires are severed the subsequent release of energy can be catastrophic with an explosion of material. The forces used are so great that it is dangerous, and skilled work is needed both to fabricate and to dismantle these elements. It is more common in the building industry to use these techniques, under controlled conditions, for the production of high-strength components, which include major structural elements and secondary components such as flooring planks.

The differentiation between prestressing and post tensioning is that prestressing involves casting concrete around wires which are under tension and the tension is released when the concrete sets; in post tensioning, wires are sleeved in tubes or wrapped and casting takes place with the wires loose, but they have room to move. They are then tensioned and grout is pumped into the space between the wires and the sleeving material. See *MBS*: *Structure and Fabric Part 2* on the principles of prestressing and post tensioning.

Storage of materials

It is important that all materials for concrete are well kept on site. Aggregates should be stored and batched on temporary hard surfaces which are sloping for drainage. Bags of cement should be kept under cover and checked for age.

Formwork

BS 5975: 1982 is the code of practice for *falsework* and reference is made back to this standard in the section 6.9 on formwork in BS 8110.

The construction of all concrete structures is limited by the engineering and building of formwork structures. Formwork is a complete subsystem of building which has to be self-supporting and also able to carry the load of wet concrete. Formwork is sometimes referred to as *shuttering* and the supporting structural framework has the alternative name of *falsework*. Where possible, elements should be designed for reuse and the parts of the formwork that will eventually require replacement are the facing materials which the concrete will be cast against. In reusing these building systems care must be taken that the construction of formwork will ensure that all joints are tight to prevent leakage of the wet material. Any leakage will affect local strength and give rise to defects such as hydration discoloration or honeycombing (see *MBS*: *Finishes*). It is ironic to think that many concrete buildings have structures limited by the extent to which timber or steel can be used as a structural material to carry the weight of wet concrete prior to hardening.

There is now a range of systems from which to choose, made from steel as well as timber, with an addition of facing materials in plastic and rubber. According to the scale of the work, decisions will have to be made on how concrete shapes can be designed to reuse formwork and not incur unnecessary waste in timber. Many architects will specify formwork by using standard clauses such as 6.9.1 of BS 8110 which will ensure stability but be neglectful of surface finishes which might affect the original specification for formwork. As formwork often accounts for 50% of the cost of construction for making a concrete frame, the repercussions of making poor decisions are obvious. There are better techniques available which now make the use of concrete competitive with steel, and there are advances in detailing concrete which optimize on the repetition of shapes for formwork by the simplification of construction details. If prime concerns are economy and speed in construction then all detailing must be looked at carefully. For example, shearheads at the top of columns would be advocated in preference to expression by the use of mushroom capitals which would need careful detailing and purpose-made formwork.

Figure 3.12 Formwork in Tokyo (1990): (top) showing the tie bars in place attached to the plywood shuttering and the walling reinforcement being placed; (middle) the outside face of the formwork showing the holders for the scaffolding poles which finally stabilize the form (bottom).

Release agents

Release agents have traditionally been used as coatings to formwork to prevent adhesion of the concrete. These

Figure 3.13 Ark Dental Clinic in Kyoto 1983 by Shin Takamatsu: (top) the building is next to a railway station and the repetitive chimney elements act as external light fittings, bringing daylight into the building; note the careful formwork patterning (bottom left) and the staining from the parapet detail, also the segregation discoloration from delay between pours on the side of the building (bottom right). Shin Takamatsu says he used hard materials 'because in Kyoto there are many wooden houses, houses with white walls, houses with stuccoed walls. They are all very mild, like human skin'.

include oils, emulsions, chemical release agents and waxes. If coatings to formwork are chosen carefully, i.e. polyurethane finishes to plywood, laquers and plastic coatings, they may not be needed. Checks should be made on the components of release agents as some can carry health hazards by causing skin irritations, and have a degree of toxicity if taken internally. Manufacturers' instructions should be followed and protective clothing worn if recommended.

Climbing formwork is now standard practice for the building of core walls in large buildings and if this kind of system is augmented by the use of precast components then the management and construction of concrete buildings can be as efficient as that of steel. *Flying formwork* which is the crane-assisted movement of steel and aluminium trusses with plywood decks, has speeded up *in situ* concrete construction so that whole floors can be cast within a four-day cycle. For further details see *Advances in concrete construction technology*, D F H Bennett, British Cement Association, 1989.

Concrete as a cast material As concrete is a cast material it is important for architects and designers to understand that they can influence the total appearance by a much more rigorous choice of formwork systems or materials against which the concrete is cast. People have been nervous about choosing fair-faced finishes as their weathering can be poor, and staining from poor detailing is often enriched by the growth of algae on heavily textured surfaces where water can lie.

Figure 3.14 The Pharoah Dental Clinic, Kyoto, by Shin Takamatsu, showing another design for bringing daylight into the building: (left) the windows use a washing machine door motif; (top middle) the jointing in the metal cladding shows a leaching of calcium carbonate from the concrete behind it which is washing over the surface; (bottom right) window detail; (top right) a better detail used by Shin Takamatsu is in the embedding of solid pieces of stainless steel in the Octagon Building.

Finishes have their own periods of fashion. In the early 1920s and 1930s Corbusier used honestly expressed board marked *finishes*. In the middle of the twentieth century there was a move towards exposed aggregate finishes and the brutalist finishes of the 1960s often required tooling or bush hammering after the basic shape was cast. One strongly featured vertical ribbed finish was often referred to as the *Elephant House* finish, after Hugh Casson's work in Regents Park Zoo. Currently in the 1990s there is an interest in smooth finishes after the work of several Japanese architects, such as Tadao Ando or Shin Takamatsu, and a greater interest in the materials used for formwork. This might involve casting against metal to achieve smooth marble like finishes. Lloyds of London by Richard Rogers had 20 sample panels cast before it was agreed that the most acceptable formwork was plastic faced plywood.

There is also no point in specifying render finishes if it is at all possible to specify concrete with the right finish. There are definite patterns in concrete today that reflect the formwork techniques used. Plywood panels with a system of tie bars for stability will produce a pattern of rectangular joints where the panels abutt, and holes where cones at the end of tie bars are expressed. These patternings, if expressed on elevations, must be worked out with the knowledge of the right systems of formwork being used, and these will have to be designed to satisfy retention of the wet concrete in the first place.

Main types of formwork

Curved walls or round columns are not a problem but materials have to be chosen carefully. It is ideal to curve plywood for walls but the tighter the curve the thinner the ply, until perhaps a shape can only be built up in vertical pieces 300–600 mm in width or around a column in 20 mm sections.

If fair-faced concrete is to be used, the details for weathering must be very carefully looked at. Water must

Figure 3.15 Peckham Health Centre 1936 by Owen Williams: (left) formwork enables this kind of profiled mushroom column to be made. Structurally efficient in using a flat slab construction, but time-consuming in terms of building formwork, this system allows for the omission of edge beams and gives a continuous soffit to the outside; (right) staircase detail where the balustrade has been cast integrally.

Table 3.11 Striking times for formwork

Type of formwork	Minimum period before striking for given surface temperature of concrete		
	16°C and above	7°C	t (any temperature between 0°C and 25°C)
Vertical surfaces to columns, walls and large beams	12 hours	18 hours	$\frac{300}{t+10}$ hours
Soffits to slabs	4 days	6 days	$\frac{100}{t+10}$ days
Soffits to beams and props to slabs	10 days	15 days	$\frac{250}{t+10}$ days
Props to beams	14 days	21 days	$\frac{360}{t+10}$ days

Source: Table 6.6 BS 8110 Part 1:1985

be shed quickly and not allowed to lie on surfaces, particularly on ramps and staircases, and recessed wall details should incorporate proper drips. All the traditional details around windows, including sills and roof details around eaves and gables, should be looked at so that water is not encouraged to flow back and down the building, thus staining will be kept to the minimum.

Striking times

As the concrete hardens it has to remain fully supported until it reaches a cube strength (measured from cubes kept next to the concrete) of twice the stress to which the concrete will be subjected when the formwork is struck. This will be the dead load of the structure, and live loadings from building operatives. See Table 3.11 for details of striking times.

As the concrete is hardening due to an exothermic chemical reaction, the rate of hardening is temperature dependent and affects the timed removal of formwork, which will more than double in winter conditions.

Case study — chapel of the Ikeda, Osaka, Japan

This is a small building which serves as a church with community facilities, including the unusual feature of a play area for children on a landscaped roof. The building is set into the side of a steeply sloping site. The construction is of reinforced concrete, fair-faced, and although the building is now at least five years old, the concrete looks reasonable and provides a permanent finish as well as being the main structure for the external wall which is inset between the column and beam system supporting the roof.

The cement is equivalent to ordinary Portland cement with a strength of 23.54 N/mm^2. No special jointing techniques were used and the formwork was plywood with a polyurethane coating with no release agents. Normal slump tests were carried out on site to ensure consistency and quality, and sample cubes were taken every 150 m^3. With this type of design, where there are large areas of concrete, it is critical to ensure that the construction sequence is planned to allow for continual supplies of ready mixed concrete to ensure consistency. The vibration of the mix must be controlled to minimize differential patterning between pours which can show up as a number of inconsistencies which include colour differentiation and sometimes honeycombing.

In a design such as this, control of the formwork is critical. Joint design has to be considered in relation to the formwork design. At this point architects should be specifying the actual material and finish of the formwork, and ensuring that formwork joints are an integral part of the design. If ties are used for self-supporting formwork

Figure 3.16 Chapel of the Ikeda, Osaka, Japan by Takenaka Corporation acting as architects and contractors.

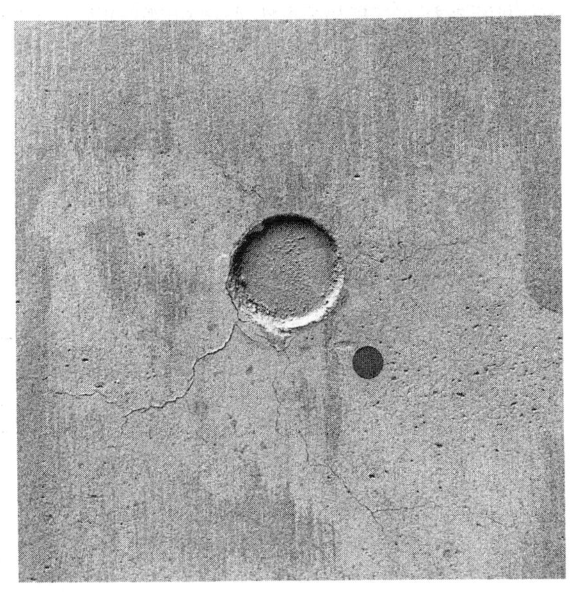

Figure 3.17 Typical tie bar end connection showing fine crack generation. Dot size is 8 mm.

then the centres of the bars will be expressed on the external face of the building, and the positions marked either by the holes left from the extraction of cones or by specially shaped plugs which fit into the holes left by the cones. There is a tendency for cracks to be generated around the positions of the tie bar ends. Reasons for this are conjectural. However, contractors should ensure that formwork components are of high quality and that the plastic cones are not made from a polymer susceptible to a high rate of water absorption and consequent expansion. This could cause swelling of the polymer, imposing pressure on the concrete at a point where it would be naturally shrinking around the cone. Both mechanisms would place the hardening concrete in a position of stress and be a possible cause of fracture initiation.

3.10 Concrete admixtures

Standard admixtures

Concrete admixtures are an important part of making concrete and the following guide owes a great deal to the clear setting out of all admixtures in the Concrete Society's Technical Report 18c published in 1980; The Concrete Society's details have been updated by reference to current technology.

Admixtures alter the properties of concrete and are additions other than water, aggregates or Portland cement. They improve workability and compaction, prevent the segregation of aggregates and consequent bleeding of cement, prevent plastic cracking by the control of shrinkage, and generally improve durability. They cannot improve the quality of an initially poor concrete mix but they can optimize its properties and exert a great deal of control during the hardening process. They can be either soluble or insoluble additives. Soluble additives would include plasticizers and pigments, and insoluble additives include retarders, air entraining agents, super plasticizers, accelerators and integral waterproofers. *Active* ingredients will work by inducing a chemical reaction such as accelerators, retarders and waterproofers whereas *interactive* ingredients will work by not chemically changing the form of cement, but use mechanisms such as the residual negative charge present in aggregates which can then attract positively charged air molecules. These include plasticizers, super plasticizers and air entraining agents. The attachment of air molecules gives the properties of surfactants to aggregates, helping to control the frictional resistance of aggregates. Completely passive admixtures include viscosity modifiers and pigments. The various types of admixtures are given below.

Accelerators: BS 5075 Part 1 They reduce the setting and hardening times and are based on calcium chloride or calcium formate. Calcium chloride is now restricted from use in reinforced or prestressed concrete due to chloride ions as initiators in the corrosion of steel reinforcement and as contributors to the shrinkage of concrete. BS 8110 should be consulted with regard to chloride ion content in different types of concrete. Accelerators have a negligible effect on long-term strength but can improve early strength by at least 25% after one day. Plasticizers can also be used to reduce water demand and accelerate early strength gain of the concrete.

Retarding mixtures: BS 5075 Part 1 These work by reducing the temperature and hence the rate of reaction at which the concrete will set. They delay the setting time normally by one to three hours but can be specified to delay setting for two or three days. This may be needed for working on surface finishes. The main chemicals are salts of lignosulphonic acids, salts of hydroxycarboxylic acids, low molecular weight polysaccharides, salts of boric acid, or salts of phosphoric acid. They compensate for concreting in hot weather and allow for a better junction in poured concrete to be made when large quantities are being placed.

Accelerated water reducing agents: BS 5075 Part 1 These increase workability, reduce the temperature rise of the mix and improve the pumpability of the mix. These all help in placing the concrete in situations where there may be complex shapes and improve the compressive strength of the concrete by controlling the water:cement ratio.

Retarding water reducing agents: BS 5075 Part 1 These reduce the temperature rise in concrete.

Air entraining mixtures: BS 5075 Part 2 They reduce bleeding and improve the workability of the concrete, minimizing particle segregation. They are organic surfactants forming bubbles 0.25–1 mm in diameter which are negatively charged, attracting themselves to positively-charged cement particles. As porosity of the mix is increased, the overall strength of the concrete is reduced. Mix design must allow for this, which can include the balancing effect of admixtures. They improve the frost resistance of concrete by the voids formed, acting as expansion chambers for ice formation in capillaries, giving a structured discontinuity in the material. Engineering frost resistance is very important where concrete is exposed to constant saturation.

Chemicals used for air entraining agents are primarily abietic and primeric acid salt (neutralized wood resins). These wood resins, which were discovered by accident as being particularly effective, are sometimes referred to as *vinsol resins* meaning *very insoluble resins*. These resins act as surfactants by lowering the surface tension of water which under mechanical agitation forms closed air bubbles which have a surrounding positive charge. These formed bubbles will then have the ability to repel each other causing bubble dispersion and an even distribution throughout the mix. The bubble size is of the order of 0.2 microns with a spacing that may be several hundred microns. It is standard practice to control the spacing to 200 microns in the USA, where there can be severe drops of as much as 12°C in temperature. ENV 206 specifies bubble spacing of this order as a directive in the UK although the freezing conditions are not so extreme. Use is also made of fatty acids and salts of fatty acids, alkyl aryl sulphonates, alkyl sulphates and phenol ethoxylates for air entraining agents.

Super plasticizers: BS 5075 Part 3 These accelerate early gain of strength and increase the workability by greater control of the water:cement ratio which increases the ultimate strength of the mix by as much as one third. They can also reduce the temperature rise and generally improve pumpability. They are all long chain molecules and often one of the following:

Sulphonated melamine formaldehyde condensates
Sulphonated naphthalene formaldehyde condensates
Modified lignosulphonates.

Pigments: BS 1014 Before using pigments, efforts should made to look carefully at the choice of natural aggregates and cements as there is great variety in the natural materials available. Colours originate from the use of iron oxides which can be natural or synthetic. Pigments are highly processed materials and are very expensive. Their use in concrete can increase costs by as much as 400%. Colour conditioners may need to be used which act as vehicles carrying the pigment but have other properties as admixtures. By increasing the workability of the mix they reduce the water content and allow for a more even distribution of the pigment.

Pigments are added to either white or ordinary Portland cement. They should account for 5–10% of the mix to be effective. The cement content should be increased by the same amount to avoid any reduction in strength. Pigments are covered in BS 1014.

Stains can be used as an alternative to colouring concrete. They are metallic salts in a water medium which react with the concrete, penetrating only 2 mm to give a smaller ranger of softer colours; greens, reddish browns and yellow ochre. Black can also be produced. They are usually applied one month after the concrete has set.

Whatever means are chosen to colour concrete, in protected situations the finished material can also be waxed to give a deeper colour rendering.

It is preferable to use pigments to control the body colour of concrete rather than relying on surface applications and they may be used initially to control cement colour. For example, white pigments may be added to grey Portland cement in controlled amounts, giving less colour variation than by relying on changeable supplies of white Portland cements. This is common practice in pre-cast concrete manufacture. The cement has to be specified simultaneously as the control of cement and pigment can be interdependent. If light or pastel shades are wanted then white Portland cement will have to be specified so that pigments will give a full colour rendering and not be altered by the cement colour.

Most pigments used are derived from inorganic minerals including red, yellow, brown and black iron oxides, black manganese oxide, blue cobalt oxide and green chromium oxide. The reason for using iron oxides as the most common pigments is that they are known to be stable within the alkaline medium of cement, are lightfast and weather resistant. Colour control is obtained by pigment shape and iron oxide reds will vary from brick red to violet red, depending on whether the pigment shape is spherical or cubic. Yellow iron pigments are needle-like in character and shades are varied by the aspect ratio, i.e. length to width. Blue cobalt oxide can affect the setting time of cement, and copper compounds should be specified in preference. Carbon black may be used but BS 1014 should be checked for the permissibility of other organic dyes. Carbon black can leach out with time and can produce flash setting. Organic dyes can degrade in alkaline environments. As a rough guide using pigments up to 1% by weight will produce tinting, up to 5% will give definable colour and between 5 and 7% yields a deep shade of that colour. If the hue chosen is a light body colour such as yellow, then up to 10% by weight of pigment will have to be added.

The workability of concrete is reduced if pigments are used, and wetting agents may have to be introduced. Reduction in compressive strength may occur with particular pigments, such as the use of a yellow pigment at a concentration of over 6% in the mix. Good control of the formwork is required and mould release agents recommended should be emulsion creams or chemical release agents, or other systems compatible with the pigments used.[13]

Other admixtures

Bonding These are organic polymer emulsions such as polyvinyl acetates which improve adhesion. Before using adhesion systems on concrete, the surfaces to be joined have to be thoroughly prepared. This situation arises in replacing damaged or contaminated concrete with a new mix and the surface conditions needed should be the same as if a cementitious mix is to be applied directly. After damaged concrete is removed, surfaces should be altered by mechanical shot blasting or sand-blasting. Bush hammering will fracture the surface using a multipointed tool and a scarifier will roughen the surface by scraping with a tungsten carbide or diamond-tipped surface. Surface preparation is characterized as *light blasting* (which can include acid etching) and removes laitance and exposes the surface of fine aggregates, *heavy blasting* which exposes the surface of coarse aggregates, normally of 3 mm amplitude which starts to remove contaminated surfaces, or 6 mm amplitude which achieves a greater depth of removal, and *fracture plane* which will fracture coarse aggregate. From heavy blasting upwards the surface should be cleaned using high-pressure water. Additional sand-blasting may be required to prepare corroded surfaces of reinforcement.[14]

Waterproofing and applied coatings Watertightness is the ability of concrete as a material to hold back water without leakage. This has to be distinguished from permeability which is the migration of water, or another liquid phase or a gas phase, through concrete under pressure. Admixtures can reduce the permeability of these various phases through concrete and inhibit some of the more damaging effects of migration of other phases, particularly the passage of chloride ions from de-icing salts or sulphate ions from groundwater. The first requirement to reduce water penetration and permeability generally, is to control the water:cement ratio. If this can be reduced to less than 0.3 the concrete can be regarded as impermeable, which would be needed for conditions of severe exposure or where a high specification is needed. A water:cement ratio of 0.4 should be adequate for most situations requiring water tightness.[15]

In the repair of concrete structures where carbonated concrete has been removed and replaced, it is common to apply coatings to reduce the permeability of concrete to carbon dioxide and oxygen which will assist in the prevention of corrosion of steel. There is a dilemma in specifying coats which must be vapour permeable, as any moisture left in the substrate can expand and cause blistering of the protective coating if it cannot escape, and yet the coating must still resist the passage of oxygen and carbon dioxide. Most coatings perform satisfactorily but their performance should be checked by inspecting applications which have some years of service. Corrosion that may take place because of chlorides in the mix can be inhibited by coatings which will stop access of water and oxygen needed for the corrosion process and so extend the useful life of concrete.

Concrete surfaces should be prepared properly (see above) using sand-blasting followed by high-pressure water jets. A range of proprietary products can be used such as chlorinated rubbers, polyvinyl copolymers (particularly acrylic styrene and terpolymers), reactive, solvent-based and water-based acrylics, epoxy resins and bituminous products. If treatments are used that do not materially alter the face of the concrete by a surface film of at least 250 microns they are unlikely to give long-term protection against carbonation. These come under the categories of penetrating sealers. Those regarded as water repellents are the least successful. Acrylic resins, silane resins, epoxies and polyurethanes give better durability. Coatings have other uses in providing dust-free surfaces for clean environments and to facilitate cleaning in food handling or the pharmaceutical industries. They can also prove effective against chemical attack in industry and contamination from nuclear industries.[16] Slurries may be used which combine acrylic styrene polymers and cement where the cement will act as a filler within a polymer matrix on hardening. Slurries are usually applied in two to four coats with an overall thickness of between 2 and 4 mm, and bridge over cracks, providing waterproofing and an impermeable coating to carbon dioxide and oxygen, but allowing water vapour diffusion. Applications of slurries are used for engineering structures such as bridges or for general repairs.[17]

Toughness The extensibility of concrete can be improved by incorporating at least 10% latex admixtures relative to the cement content. These can reduce the water:cement ratio to 0.33 and improve the abrasion resistance of concrete. This is useful in bridge decks and on the steel decks of ships where the loading of structures is more dynamic. Organic polymers used include:

Styrene butadiene
Styrene–butyl acrylate
Methyl methacrylate (MMA)
Vinyl acetate

The temperature at which latex will cure is critical. At 2°C there will be no adhesion to steel, this improves at 20°C. Stabilizers have to be added to latex to prevent a lumpy coagulation with cement. This is due to calcium ions which can destabilize the solution. Manufacturers' advice should be sought for different applications.

Damp-proofing and waterproofing Waterproof concrete mixtures should rely initially on a dense concrete mix with no gap grading. Damp-proofing admixtures reduce the rate of water absorption but do not reduce the permeability of concrete. They include stearates, oleates and other organic chemicals in a latex form.

3.11 Mixing of concrete

Concrete can be mixed on site or brought to site as ready mix from works where it is mixed in large quantities and distributed to sites. Ready mixed concrete now accounts for some 80% of all concrete used in the UK. It is controlled by BS 5328 and also by a Quality Assurance Scheme operated by an independent body with certain ready mix plants as members. The use of ready mix concrete requires good organizational site methods and can be difficult to programme in urban sites when the delivery of loads is held up by traffic, therefore the length of the working day on site may have to increase to compensate for any delays. Site access conditions are important to check. As each ready mix truck weighs 24 tonnes, access roads must be able to take the load adequately and provide a turning circle of 18 m. If the concrete is to be placed close to truck access it can be placed by chute, if not, skips carried by crane or pumps will have to be used.

Mixing directly on site will only happen for small jobs or those which are so large, as in the case of civil engineering contracts for bridges, reservoirs or motorways, that large-scale mixing is the only solution. Some large-scale architectural projects may need the production of pre-cast elements on site.

Ready mix concrete may not be appropriate if special finishes are to be achieved and in all cases sample panels should be made for fair-faced work to ensure that there is a visual check on the final result. Most pre-cast elements are produced under workshop-controlled conditions and then transported to site. Once concrete is mixed, and ready for placing, the time lapse should not exceed 30 minutes.

All machines used for mixing concrete have to be cleaned every day, usually with water and loose aggregates.

Placing of Concrete

The making of concrete is a large-scale casting operation where the strength of the material has been calculated and the mix design is critical.

The whole process of placing concrete has to be under careful control to ensure that the mix does not suffer in quality during transportation and placing in formwork. To ensure a homogeneous material the process of casting should be continuous and the mixture should undergo continuous evaluation to ensure consistency. On no account should the mixture be allowed to free fall otherwise heavy particles will finish up at the bottom; a phenomena known as density separation. As the particle sizes segregate out, the number of voids in the mix will also increase as homogeneity is lost.

There are also different workability requirements. Filling vertical walls will require a greater workability than if placing the concrete for slabs. When ready mix lorries are used, placing should be programmed so that there is always a lorry waiting to complete a pour. Otherwise segregation can occur and a distinctive patterning can develop between pours which will become more apparent on the finished surface after time.

Pumping Concrete can be pumped through tubes which can be 75 or 100 mm in diameter with a normal maximum length for placing of 300 m or at most 600 mm with a vertical head of 50 m, although special pumping will give height reaches of 300 m. The mix used for pumping should have high workability and site testing should show a 75 mm slump. It is important that the mixture flows from the tube directly into the formwork and is moved along, being raised as the mixture level rises. Special *tremie* pipes have to be used if concrete is to be dropped to a lower level to avoid segregation of aggregate.

Compaction After placing the concrete it has to be compacted by removing voids. This can be achieved by overfilling and physically tamping the concrete into place, or by using mechanical vibration. *Poker vibrators* are used which allow air bubbles to rise to the surface with a cement-rich thin film. When this activity stops the poker can be moved along usually at intervals of between 300 and 500 mm. When pre-cast elements are made, the concrete is poured into forms which are vibrated as a whole on tables. Surface vibrators are only used for concrete which has a maximum depth of 150 mm for floors or roads.

There is an approximate loss of strength of 5% for every 1% of air in the mix. For a concrete mix to be durable it must be dense.

Construction joints It is inevitable there will be joints in concrete. These may be movement joints to allow expansion and contraction, or construction joints. Construction joints separate casting at different times and are commonly referred to as *daywork joints*. The surface for construction joints has to be prepared by removing the cement-rich topping and exposing the aggregate ready for bonding to the next pour. If the surface is green this can be done by straightforward water spraying and brushing. If the surface has hardened, techniques may need a greater amount of mechanical treatment, described in section 2.3: Bonding. It is not recommended to use retarders in this situation as their total removal prior to the next pour is difficult to supervise and they are more likely to cause poor bonding. As construction joints are inevitable, formwork should be designed with clear separation details.

Both kinds of joint may need to incorporate water bars to prevent water penetration, but in the case of movement joints reinforcement will have to be discontinuous. The joint is then filled with a compressible strip, finished with a sealing compound for weatherproofing.

Concreting in hot and cold weather The curing of concrete has to be controlled to reach optimum strength no matter what the weather conditions might be. Admixtures can be used to retard the concrete in warm conditions or to accelerate it in cold conditions. As water is essential for the hydration of concrete the surface has to be kept damp and might be covered with polythene to prevent any evaporation. In warm weather, surfaces can be used that are solar reflective. In cold conditions the concrete might be covered over with a polythene tent with a 500 mm air gap to provide some insulation. The actual mix may use warm water and insulation with blown warm air. The formwork techniques may change in response, using timber or 19 mm plywood panels to provide some insulation. Concrete should not be placed if the temperature is over 30°C. Any possibility of thermal cracking will be accelerated if the ambient temperature rises.

Testing

Testing is a necessary part of the process in using concrete. It is such a variable material that checks are needed to validate the quality of the mix on site, and cube tests have to be taken at regular intervals to ensure that the expected strength of the mix is being achieved for the loading conditions calculated.

Testing in concrete involves a range of measures which first involves the sampling of all materials and checking the sizes of aggregates delivered. The individual material constituents of cement, aggregates and water have the following standards which would ensure comparable evaluation by different laboratories:

Cement: BS 4550 The testing of cement is always carried out under laboratory conditions which have controlled humidity and temperature.

Aggregates: BS 812 This determines the amount of fine clay and dust in an aggregate sample. If a field settling test shows that more than 10% of the fine aggregate is silt, further controlled tests in a laboratory should be carried out. This standard also sets out methods for determining the compressive strength.

Water: BS 3148 It is important that water is chemically pure and free from any impurities that would affect the setting of the mix. Mains water should be of a sufficiently high purity to meet this standard.

When a concrete mix is tested, samples should be taken, preferably from a flowing stream of concrete while it is being placed and not when stationary. The major test carried out on site is the slump test which helps to determine the water:cement ratio (often the greatest variable on site) and the workability.

Mix testing

Testing can be divided immediately into two categories: *on-site testing* and *off-site testing*. All tests are set out in the British Standards and, for guidance, reference should be made to BS 1881: 1970–71 which is in six parts covering the testing of fresh as well as cured specimens. BS 4408 covers the non-destructive testing procedures for concrete *in situ*.

On-site testing The most important factor is the consistency of mix on site. In wall constructions different mixes can be very apparent and should be avoided in fair-faced finishes. Slump tests provide evidence of the workability of mixes, and the megaphone-like conical mould has to be filled with concrete in four stages each of which are each rodded carefully with 25 strokes of the tamping rod. After the mould is removed, the height of slump is recorded. There are three categories: a *true* slump, *shear* slump or a *collapse*.

The compacting factor test is more accurate as it is imitative of the way concrete is placed and it is a comparative test. The mix is placed in the top cone, without being compacted, and allowed to fall to the next cone and then to the cylinder below with an excess being trimmed off by working two floats simultaneously across the mix. The weight of the concrete in the cylinder is then determined and compared with the weight of the same mix fully compacted in the cylinder.

Off-site testing

Air content test This is used if air-entrained mixes are specified and is carried out on compacted concrete samples with a specialist air meter to BS 1881 Part 106: 1983 *Method for determination of air content of fresh concrete*.

Concrete cube tests These are used to check the compressive strength of the concrete being used and should be made and cured to BS 1881 Parts 108 and 111. Concrete put in the cubes should be fully compacted. 150 mm cubes should be filled in three layers compacted with 35 strokes, and 100 mm cubes compacted with 25 strokes in two layers. Each layer has to be compacted using a steel bar which is a standard 380 mm long, weighing 1.8 kg and of a square cross-section of 25 mm^2, rounded at the top end to form a handle. Alternatively the cube can be subjected to vibration from a pneumatic hammer. The cube should then have the surface trowelled smooth, level with the top of the mould. Cubes have to be properly cured, wrapped with polythene to prevent moisture loss and kept at a constant temperature of 20 ± 5°C if they are to be tested after seven days and ±2°C if they are to be tested after 24 hours.

3.12 Glass reinforced cement

See *Guide specification for cladding*, published by the Glass Fibre Reinforced Cement Association (GRCA in the UK) for more details.

Glass reinforced cement (GRC) has developed as a new composite material in the 1970s, with glass fibre incorporated into a cement and sand mix, as a fibre composite that can increase the tensile strength of ordinary cement. The mix can be modified with a polymer to give additional strength in tension. This gives scope for thinner sections of material which can span and become structural panels, depending on their configuration, usually referred to as *thin-walled panels* between 10 and 12 mm in thickness. The Glass Fibre Reinforced Cement Association was formally constituted in 1975 and progressed work in this field after development by the Building Research Establishment. Alkali-resistant glass had to be developed by Pilkingtons as the strength of a fibre composite is dependent on the purity and strength of the individual fibres. The alkaline environment of cement can cause surface defects on ordinary glass which make it notch sensitive and likely to fracture. The fraction of glass fibre used varies between 3 and 6% by weight. Individual fibres vary between 12 and 38 mm in length. The aggregate:cement ratio is from 0.25 upwards and the aggregate is a fine sand with a particle size between 150 microns and 1 mm. Polymers may be incorporated to improve the properties of the material.

GRC can be made by casting or spraying or pressing the mixture into a mould. The stiffness of panels or components can be enhanced by careful design of the geometry of the component, using *folded* sections or introducing tubular stiffener ribs in GRC or steel.

Using a stud-frame technique developed in the USA, structural panels can be used for cladding or permanent formwork. The face can be cast to reproduce any finish required and this composite is a high strength material that can be used in some situations for restoration work where it is not possible to use original materials. Care should be taken in restoring external features, e.g. cornices on a stone building where differential weathering should be allowed for. The face of a GRC panel can also be clad with a variety of finishes including stone. The design of a panel should be undertaken in conjunction with an approved GRC sub-contractor.

As an alternative to flat single panels, sandwich panels can be used with insulating cores. There are disadvantages to this method of construction as the thin skin elements are responsive to changes in moisture and temperature and any differential between two skins that are firmly bonded can lead to flexure and tensile cracking. Panel design should allow for independent movement of the outside skin. An alternative method of construction will have a bearing skin of prefabricated concrete carrying a sandwich of insulation and outer skin.

GRC is also used for pipes providing a strong light and waterproof conduit that is highly resistant to chemical attack and, as an applied render mix to ordinary concrete blocks, provides a tough finish that is highly resistant to cracking. In a slab form it can provide hardwearing wall surfaces for tennis exercise walls or skateboard surfaces which can then be made into complex shapes.[18]

Generally GRC is a strong lightweight composite which can be combined with slabs of insulation to form a large-scale composite cladding material. As a material it has the advantage of performing to Class O in terms of combustibility, with a fire resistance between one and two hours. Due to its dense mix, which limits gas permeability, it has some advantages over ordinary concrete with greater resistance to carbonation and chloride penetration. This demonstrates that GRC gives potentially greater protection to reinforcement than concrete and is ideal to use as permanent formwork for high-level protection (*GRCA News*, September 1989). Permanent bonding between formwork and the concrete core should be achieved to maximize the fire resistance for steel reinforcement and protection against carbonation. Because of its brittle nature, or poor behaviour in bending, GRC should only be used for formwork for vertical cladding where it can satisfactorily contain horizontal pressures.

Where GRC is used as the finishing material, maintenance by cleaning and resealing approximately every ten years is advisable, especially in urban conditions. BRE Digest 331, April 1988, on GRC should be referred to for detailed results on its performance with regard to its general properties and testing to determine permeability, behaviour under fire, fatigue and freeze–thaw behaviour. The Digest also outlines the changes in performance as a result of using different mixes, particularly if OPC is replaced by pulverized fuel ash or granulated blast-furnace slag.

3.13 Health

Lubricants Under this category some mineral oils used as mould release agents in formwork can cause dermatitis or acne, and may be a cause of skin cancer. In an aerosol form they can cause respiratory damage. Good ventilation is necessary and protective clothing which is impervious to oil should be used.

Silica Silica is used in finishing treatments for masonry and concrete and is a cause of silicosis and other respiratory complaints. Surfaces treatments which use blasting should always be carried out in conjunction with water or oil, and good ventilation arranged.

Cement Cement is listed in *The control of substances hazardous to health in the construction industry.*[19] As dust, it can cause dermatitis from chromate substances, and the lime content can cause skin burns and affect the nose, mouth and eyes. Close working by bricklayers or plasterers in construction is hazardous. Waterproof boots and gloves are advised to be worn when mixing materials and the use of barrier creams is encouraged. Gypsum can give the same level of irritation (see Health Hazard Sheet 5 of the COSHH document).

Site preparation, foundation work and health

Workers involved in the preparation of sites and foundation work will encounter other hazards in the form of contaminants from the ground. These active substances in the soil might cause tetanus or hepatitis. In addition, the previous use of sites can yield toxic metals. Different industries generate varying pollutants and pre-knowledge of the site is advantageous. It is responsible to find out the previous use of sites and to examine the soil for possible pollutants so that workers can take precautions. Previous gas work sites will be heavily contaminated and former tanneries or hospitals may have heavy chemical pollution with such toxic substances as ferrocyanides, cadmium, arsenic and phenols. Toxicity can be transferred by inhalation and ingestion, as well as local skin contact which can be a cause of the absorption of chemicals. Rat infestation can transfer Weils disease. As an extreme example, radioactive contamination was found on the site of a former factory in the East End of London which prepared luminous dials for aircraft during the Second World War. Census reports are a good source of information for previous site use. There is also a need not only to identify main pollutants but also to monitor the site once pollutants are discovered. National and local registers of contaminated land are now being set up which will help people to determine major issues that have to be tackled.

References

1 CRATerre Editions Parentheses 1989 Traite de construction en terre. In Hugo Houben and Hubert Gullard (eds) *L'encyclopaedia de la construction en terre, vol. 1.*

2 ENTP Laboratorie Geomateriex (Director: Myriam Olivier), Rue Maurice Audin, 69518 Vaulx en Valin, Cedex; Tel. 72 04 70 70, Fax 72 04 62 54.

3 John Ashhurst 1983 *Mortar plasters and renders in conservation* Eccleciastical Architects and Surveyors Association.

4 BRE *Annual Review* 1989.

5 G Somerville 1985 *Engineering aspects of alkali-silica reaction* C & CA.

6 D W Hobbs 1986 Alkali-silica reaction in concrete. *The Strucural Engineer* **64A** (12) December.

7 BRE *Digest 330*, March 1988 (replaces Digest 258).

8 S Chandra and L Berntsson 1988 Deterioration of concrete in swimming pools in the south of Sweden. *ACI Materials Journal* November/December: pp. 489–94.

9 X Zhang and G W Groves 1989 Microstructural studies in alkali-silica reaction products. *Materials Science and Technology* **5** July: pp. 714–17.

10 B Zech and M Setzer 1989 The dynamic modulus of hardened cement paste: Part 2 Ice formation, drying and pore size. *Materials and Structure* **22**: pp. 125–32.

11 A Bochenek and G Prokopski 1989 The investigation of aggregate grain size on fracture, toughness of ordinary concrete structures. *International Journal of Fracture* **41**: pp. 197–205.

12 From R D Anchor, A W Hill and B P Hughes *Handbook on BS 5337:1976.*

13 C J Lynsdale and J G Cabrera 1989 Coloured concrete: A state of the art review. *Concrete* **23** (7) August: pp. 29–34.

14 Miles Murray 1989 Surface preparation for adhesives. *Concrete International* September: pp. 69–71.

15 Permeability of concrete. *Concrete Construction* 1989 October: pp. 870–72.

16 John Shaw 1989 Water repellant systems. *Construction Repair* April: pp. 8–9.

17 A Volkwein, R Petri and R Springenschmid 1988 Protecting concrete by flexible waterproofing slurries. *Betonwerk+Fertigteil+Technik* August: pp. 30–36.

18 Prof. J N J A Vambersky 1989 Design rules for architectural panels in glass-fibre-reinforced cement. *Concrete Precasting Plant and Technology* (7): pp. 24–33.

19 Published by the Health and Safety Executive 1989.

4 Masonry Construction

4.1 Definition of masonry, bricks and blocks

For the structure of brickwork see *MBS: Structure and Fabric Part 1* Chapter 5 and *Structure and Fabric Part 2* Chapter 4.

Monolithic construction is used here as a descriptive term for building with materials that give the appearance of a *monolith*, as if made as one whole element. Although the whole building may be made as a structure of many small pieces, they act together in a composite manner. Monolithic constructions can be made out of concrete, bricks, blocks and stone, and they have the common structural elements of wall, column and beam which are directly load bearing. Concrete elements can also be used as frames.

Masonry is used here as a term covering brick, block and stone contruction. There are common principles in putting together walling structures although the individual units may be of a variety of materials. The British Standard for masonry construction (BS 5628) should be referred to before looking at the individual standards and codes of practice for units of different materials.

Bricks and *blocks* are conveniently sized building elements. Bricks and blocks can be made from a variety of materials, their size gives them the categorization of being either a brick or a block. However, it is common to infer that bricks are mostly clay, as this describes 90% of the bricks used although there are calcium silicate bricks (3%) and concrete bricks (approximately 8%). Although BS 3291 describes a brick as a walling unit with limitations on dimensions of 337.5 mm long, 225 mm wide and 112.5 mm high, with any other units outside those dimensions described as blocks, the real criteria for the size of a standard brick is a unit that can be conveniently handled and placed in one easy action. Bricks could even be slabs of peat, commonly used in Ireland. Bricklayers have sometime been rudely described as efficient brick-laying machines for that very reason.

Figure 4.1 This café on the Isle of Dogs exploits the small element of brick to create a pattern that is contained within the stretcher bonding of the wall.

> A very large brick is inconvenient for an ordinary man to grasp, and a heavy brick fatigues the bricklayer, who has to lift it when wet and lay it with one hand.[1]

Bricks and blocks cannot be thought of in isolation as materials but must have a specification that also includes the correct mortar type appropriate to their use in terms of overall wall compressive strength required, degree of exposure and durability requirements. There can also be massive differences between the theoretical performance of a wall and its practical performance. If a mortar is too hard, or there is inadequate bonding with bricks that have poor absorption, there will be fine cracks which provide a pathway for driving rain. Most penetration of water through external walls is through fine cracks and perpends, and possibly movement joints. On buildings where a high performance is required, for example, in high rise building, testing should be carried out on sample panels under controlled conditions. The results from these tests will improve quality control on site. It is particularly important for testing to be programmed for works which required panelled wall systems within structures where there are junctions between frames and brickwork, and where a high level of skill is needed in the application of sealants.

Masonry is described in BS 5628 as an assemblage of structural units which can be *in situ* or prefabricated, with the units put together with mortar or grout. It describes brick, block and stone masonry. Part 3, section 3 of this standard describes the common design problems inherent in choosing small structural units to build from. They include:

Structural stability This may be dependent on geometric shape of the whole walling/columnar system as well as the bonding system employed.

Movement joints These recognize the natural expansion and contraction of materials through adjustment to changing moisture and thermal conditions. There will also be minor adjustments in response to loading and settlement. There is a general rule that in unrestrained brick masonry in parapets or lightly restrained walling systems, the maximimum movement will be 1 mm per metre. The width of a movement joint in millimetres can be calculated as 30% more than the centres of joints in metres. As an example BS 5628 then states for joints at 12 m centres, the movement joints will be 16 mm wide.

Table 4.1 shows the order of movement. Where possible manufacturer's information should be used as mixes of clay and concrete vary greatly.

Adhesion There must be successful mechanical and chemical bonding between mortar and structural unit. Excessive water loss must be prevented by over absorption from the structural unit or from evaporation of moisture in warm conditions, otherwise there will be insufficient water left for the chemical reaction needed for the hardening of the mortar to take place.

Figure 4.2 Colegio de las Teresianas de Sant Gervasi de Cassoles 1988. This convent by Gaudi in Barcelona uses the principle of corbelling to create most of the openings. The separate components of wall and lintel amalgamate in one structure fulfilling both functions.

Table 4.1 Linear thermal movement of masonry units and mortar

Material	Coefficient of linear thermal expansion ($\times 10^{-6}$/K)
Fired-clay masonry units	4–8
Concrete masonry units	7–14
Calcium silicate masonry units	11–15
Mortars	11–13

Source: Table 19 BS 5628 Part 3:1985

Resistance to rain penetration This is dependent on good specifications for the masonry and mortar, detailing for maximum weather protection, well placed damp-proof systems, and the correct choice of structural units and mortar for the degree of exposures expected (see Table 4.2). An estimate of the wind driven index can be evaluated from clause 21.2 of BS 5628. Workmanship is particularly important to ensure that specifications are carried out correctly. Most water penetration is through joints. If cavities are filled with insulation it is more likely for them to be subject to rain penetration.

Durability This is linked to the correct choice of masonry and mortar, being related to the degree of exposure expected. Longevity is dependent on the general durability of the masonry which is best predicted from established usage of particular components. Response to frosty conditions can still be unpredictable despite tests and manufacturers' experience is invaluable here to guide users to well-tried masonry components.

Fire, thermal and acoustic parameters These also have to be considered and are part of the design process by either direct choice to satisfy the needs in a design or in defining a likely performance specification. Table 4.3(a) and (b) gives the notional fire resistance of walls. (Refer to Table 16 of BS 5628 1985:Part 3 for full details, and to Appendix A of Document B of the *Building Regulations*.) The process of making bricks has already subjected them to extremely high temperatures and in load bearing construction a 100 mm brick wall gives one hour's protection whereas a 200 mm wall gives two hours of fire resistance. Reference should be made to the CIBSE guide, section A3, *Thermal properties of building structures* for the calculation of U values. Brickwork has little value in its own right as an insulant and will also have to be calculated with insulation. For example, a 215 mm solid wall construction has a U value of 2.0 and a cavity construction only 1.5. Acoustic performance and sound reduction relates directly to the density of material. For example, a 102.5 mm brick wall plastered both sides weighing 220 kg/m^2 will give a sound reduction of 46 dB and a 215 mm wall of the same density plastered both sides will give 49.5 dB reduction. The code of practice takes all of these factors into consideration and discusses them at length. Major problems with regard to durability are set out in this chapter. Table 13 of BS 5628 1985:Part 3 gives recommendations with regard to the specification of different kinds of masonry, and the appropiate mortar in relation to their situation. Mortar designations and compositions are given in Tables 4.4 and 4.5.

Thermal performance of walling systems As the thermal performance of walls increases, there is a danger that the outer skin may be more vulnerable to frost attack. Construction detailing should also take care in eliminating cold bridges otherwise there will be problems of condensation. However, the density of the brickwork is important in calculating the thermal mass of a walling system with regard to heat retention, and conversely it then becomes appropriate to have dense brickwork internally with external insulation. This leads logically to the conclusion that if rendered finishes are to be used incorporating external insulation there is an opportunity to reverse the traditional

Table 4.2 Factors affecting rain penetration of cavity walls

Factor affecting rain penetration	Increasing probability of rain penetration (left to right)			
Applied external finish	Cladding	Rendering	Other (e.g. masonry paint, water repellant)	
Mortar composition	Cement:lime:sand	Cement: sand plus plasticizer or masonry:cement:sand		
Mortar joint finish and profile	Bucket handle, weathered, etc.	Flush	Recessed, tooled	Recessed, untooled
Air space (clear cavity)	Over 50 mm	50 mm	25 mm	None
Insulation	None	Partial filling with 50 mm air space	Filled with type A insulant (50 mm cavity)	Filled with Type B insulant (50 mm cavity)

Source: Table 11 BS 5628 Part 3:1985

Table 4.3(a) Notional fire resistances — load-bearing single-leaf walls

Material	Masonry unit	Type	Finish	Minimum thickness of masonry (mm) for given notional period of fire resistance						
				6h	4h	3h	2h	90 min	60 min	30 min
Fired, brick-earth clay or shale	Brick	Solid	None	200	170	170	100	100	90	90
			VG	170	100	100	90	90	90	90
		Not less than 50% solid	SC/SG	—	—	—	215	215	215	215
			VG	—	215	215	215	215	215	215
Concrete or calcium silicate	Brick	Solid	None	200	190	190	100	100	90	90
			VG	200	100	100	90	90	90	90
Concrete, class 1 aggregate	Block	Solid	None	150	150	140	100	100	90	90
			VG	150	100	100	90	90	90	90
		Other (e.g. hollow)	None	—	—	—	100	100	100	90
Aerated concrete (density 480–1200 kg/m^3)	Block	Solid	None	215	180	140	100	100	90	90
			VG	180	150	100	100	90	90	90

Source: Table 16 BS 5628 Part 3:1985

Table 4.3(b) Notional fire resistances — non-load-bearing single-leaf walls

Material	Masonry unit	Type	Finish	Minimum thickness of masonry (mm) for given notional period of fire resistance						
				6h	4h	3h	2h	90 min	60 min	30 min
Fired, brick-earth clay or shale	Brick	Solid	None	200	170	170	100	90	75	75
			VG	100	100	90	90	90	75	75
		Not less than 50% solid	SC/SG	—	—	—	215	215	215	215
			VG	—	215	215	215	215	215	215
Concrete or calcium silicate	Brick	Solid	None	200	170	170	100	90	75	75
			VG	100	100	90	90	90	75	75

Source: Table 16 BS 5628 Part 3:1985

Table 4.4 Mortar designations

Mortar designation	Type of mortars		
	Cement:sand or cement-lime-sand	Air-entrained mortars	
		Masonry cement:sand	Cement:sand with plasticizer
i	1:0 to $\frac{1}{4}$:3	—	—
ii	1:$\frac{1}{2}$:4 to 1:$\frac{1}{2}$:4$\frac{1}{2}$	1:2$\frac{1}{2}$ to 1:3$\frac{1}{2}$	1:3 to 1:4
iii	1:1:5 to 1:1:6	1:4 to 1:5	1:5 to 1:6
iv	1:2:8 to 1:2:9	1:5$\frac{1}{2}$ to 1:6$\frac{1}{2}$	1:7 to 1:8
v	1:3:10 to 1:3:12	1:6$\frac{1}{2}$ to 1:7$\frac{1}{2}$	1:8

All proportions are by volume.

Source: A basic guide to brickwork mortars July 1988 Brick Development Association

cavity construction in brickwork. It would be more efficient to use a rendered insulating block externally and a denser inner wall leaf.

In the ISO Standard 9165, *Practical thermal properties of building materials and products*, attention is drawn to the thermal conductivity of the mortar as it affects the thermal resistance of a wall. Although certain thermal standards are assumed from the combination of insulation and cavity systems, recent research has shown that measured performance of built systems falls well below that expected theoretically. This can be due to poor workmanship that gives a high degree of air infiltration through mortar joints and around openings for doors or windows, or poor placing of insulation.

Refer to *MBS: Environment & services* for the theory and practice in thermal design.

Table 4.5 Mortar compositions

Type	% proportion by dry mass		Traditional volume proportioning	Designation
Cement:sand screeds	*Portland cement*		*cement:sand*	*screed*
	20.5–25.0		1:3	a
	16.0–20.0		1:4	b
	13.0–15.5		1:5	c
Cement:sand mortars and plasters (including air-entrained)	*Portland cement*		*cement:sand*	*mortar*
	20.5–25.0		1:3	i
	16.0–25.0		1:3–1:4	ii
	11.5–16.5		1:5–1:6	iii
	8.5–12.5		1:7–1:8	iv
	8.5–11.0		1:8	v
Masonry cement:sand mortars and plasters	*masonry cement*		*masonry cement:sand*	*mortar*
	17.0–27.5		1:2½–1:2½	ii
	12.5–19.5		1:4–1:5	iii
	10.5–15.5		1:5½–1:6½	iv
	9.5–12.5		1:6½–1:7	v
Cement:lime:sand mortars and plasters	*Portland cement*	*lime* $(Ca(OH)_2)$	*cement:lime:sand*	*mortar*
	20.0–25.0	0–3.0	1:0 to ¼:3	i
	14.0–19.0	1.5–4.5	1:½:4 to 1:½:4½	ii
	11.0–15.5	3.0–7.0	1:1:5–1:1:6	iii
	7.5–10.0	4.0–8.5	1:2:8–1:2:9	iv
	5.5– 8.0	6.0–10.0	1:3:10–1:3:12	v
Gypsum:lime:sand plasters	*gypsum* $(CaSO_4\frac{1}{2}H_2O)$	*lime* $(Ca(OH)_2)$	*gypsum:lime:sand*	
	16.0–20.0		1:0:1½	
	12.0–14.5	7.5– 9.5	3:2:6	
	8.5–10.5	8.0–10.0	1:1:3	
	3.0– 4.0	9.0–11.0	1:3:9	

Source: Table 3 BS 4551:1980

4.2 Accessories for masonry construction

These may be described as damp-proof courses, ties, fixings, sealants and fillers. All should have a life compatible with the expected life of the building, and economies cannot be made on these components if it jeopardizes long-term durability.

Structural fixings for masonry

Masonry now has incorporated a number of components which include:

Ties for cavity construction and bonding
Hangers for joists
Straps for tying down roofplates
Cramps and wires for cladding
General anchorages dowels and fixings
Reinforcing mesh

Initial corrosion of these components is likely within an initially saturated environment during building. Protection is required in generally exposed situations and it should be assumed (as for all-purpose fixings) that they should survive for the life of the building. As wall ties are regarded as a main element, contributing to the structural stability of walling, they are now described in greater detail.

Wall ties

As the majority of wall construction is cavity built, walling elements also rely on the longevity of wall ties. Of the cheapest wall ties produced, galvanized butterfly wire types have proved to be inadequate in certain situations. Depending on the degree of exposure it may be advisable to specify stainless steel or phosphorus bronze ties. If metal ties corrode and decay, the negative pressure on walling from wind forces can peel away the outer wall of a cavity. Generally they should comply with BS 1243:1978, revised 1982.The first amendment states:

> Recent investigations of the conditions of ties in cavity walls have indicated the corrosion of galvanized steel in this location, specifically in the outerleaf, occurs at a higher rate than previously assumed. With vertical twist ties the consequence of corrosion will often be observed by cracking of the mortar joints, and in

extreme cases by distortion of the brickwork. For the wire tie, visible external evidence of corrosion may not be seen.

There were no special conditions with regard to type of mortar or particularly aggressive situations, such as marine environments which would have increased the likelihood of corrosion. Despite this clear warning there were no significant changes made to the use of wire ties in the standard.

In BS 5628 Part 1 vertical twist ties are recommended for optimum support in unreinforced masonry, although flexible ties are advocated for situations which have to cater for differential movement or for minimizing transfer of sound. Twist ties only are recommended for reinforced masonry in Part 2 of BS 5268.

Tables 4.6 and 4.7 give specifications for the selection and for the spacing of wall ties; the spacing is dependent on the width of the cavity and leaf thickness.

Table 4.8 shows the specification required, either for the treatment of metals in terms of their coatings or their specification as alloys. Where possible it is preferable to specify non-ferrous or stainless steel fixings.

Damp-proof courses

Damp-proof courses are used to prevent moisture passing through the building fabric, especially from areas which are vulnerable to water penetration either from the ground or other openings. There are a range of materials which can provide good performance, but they all require careful detailing to ensure there is full continuity in three dimensions. Often working drawings tend to show buildings in section and three-dimensional drawings or axonometrics should be made of critical junctions otherwise continuity of protection can be lost.

All damp-proof courses must comply with British Standards; see Table 4.9 for their general performance. Useful British Standards are given in Table 4.10.

There are also impregnation treatments for existing buildings as well as the insertion of a continuous physical barrier. For some traditional buildings, especially in the case of stone construction, advice can be sought from organizations such as the Society for the Protection of Ancient Buildings. In these cases it may be more appropiate to ensure the building is well ventilated and land drainage is looked at carefully to avoid the buildup of water at ground level.

Sealants and compressible fillers for movement joints

Movement joints and edges to panelled brickwork use a variety of materials which should be specified to British Standards.

One part polysulphide sealants	BS 5215
Two part polysulphide sealants	BS 4254
Silicone based building sealants	BS 5889
Mastics	BS 3712
Cellular polyethylene	BS 6213
Polyurethane rubbers	BS 6213
Foam rubbers	BS 6213

Refer also to table 13 in BS 4974 (which is attached as an appendix to BS 5628) for sealant details.

Table 4.6 Spacing of wall ties

Least leaf thickness (one or both) (mm)	Type of tie	Cavity width (mm)	Equivalent number of ties per square metre	Spacing of ties (mm)	
				Horizontally	Vertically
65–90	All	50–75	4.9	450	450
90 or more	See Table 4.7	50–150	2.5	900	450

Source: Table 9 BS 5628 Part 3:1985

Table 4.7 Selection of wall ties

Type of tie in BS 1243	Cavity width (mm)	Characteristics	
Vertical twist	150 or less	↑ Increasing strength	
Double triangle	75 or less		
Butterfly	75 or less		↓ Increasing flexibility and sound insulation

Source: Table 9 BS 5628 Part 3:1985

Table 4.8 Metalwork in brickwork

Category	Base material	Form	Grade and standard to be complied with	Protective measures to be carried out after fabrication
A	Hot-dip galvanized low carbon steel	Sheet	BS 2989, Z1 or Z2, coating type G 600. Minimum mass of coating 600 g/m^2 including both sides	All external cut edges to be protected using a one-pack chemical-resistant paint complying with HF1A to HF2F in part 4 of table 4H of BS 5493:1972 and modified to give adequate adhesion to the fixing
			BS 2989, Z1 or Z2, coating type G 275. Minimum mass of coating 275 g/m^2 including both sides	Coating to be applied after fabrication to the external surfaces and consisting of either: (a) bituminous solution complying with types 1 or 2 of BS 3416 and of minimum thickness 25 μm, or (b) a one-pack chemical-resistant paint complying with HF1A to HF2F in part 4 of table 4H of BS 5493:1977 and modified to give adequate adhesion to the fixing. Where the zinc is removed on internal surfaces during fabrication, e.g. by welding, further protection should be applied to these areas
B	Low carbon steel	Strip	BS 1449:Part 1:1983 (mechanical requirements in table 11 only) BS 4360 grade 43A	Post-galvanizing complying with BS 729. Minimum mass of coating 460 g/m^2 including both sides
C	Low carbon steel	Strip	BS 1449:Part 1:1983 (mechanical requirements in table 11 only) BS 4360 grade 43A	Post-galvanizing complying with BS 729. Minimum mass of coating 940 g/m^2 including both sides
D	Copper		BS 6017	
	Copper alloys		BS 2870:1980, grades listed in tables 8 and 12 BS 2873:1969, grades listed in tables 4 and 6 BS 2874:1968, grades listed in tables 6, 8 and 9 except CA 106	Material other than phosphor bronze to be formed either: (a) by bending at dull red heat and allowing to cool in still air; or (b) by cold forming and subsequently stress relief annealing at 250°C to 300°C for 30 min to 1 h. Effectiveness of stress relieving of cold formed components to be tested by the supplier using the mercurous nitrate test described in clause 11 of BS 2874:1969
	Austenic strainless steel, minimum 18/8 composition and excluding free machining specifications	Strip	BS 1449:Part 2	
		Bar/rod	BS 970:Part 1	
		Tube	BS 6323:Part 8	
		Wire	BS 1554 BS 3111:Part 2	

Source: Table 1 BS 5628 Part 3:1985

Table 4.9 Damp-proof course specifications

Material	Minimum mass (kg/m^2)	Minimum thickness (mm)	Joint treatment to prevent water moving		Liability to extrusion	Durability	Other considerations
			Upward	Downward			
A Flexible							
Load complying with BS 1178	Code no. 4	1.8	Lapped at least 100 mm	Welted	Not under pressure met in normal construction	Corrodes in contact with mortars. Protect with bitumen or bitumen paint of heavy consistency applied to the corrosion-producing surface and to both surfaces of the lead	May be easily worked to required shape but this is a slow process
Copper complying with BS 2870 grades C104 or C106 in the O condition	Approx. 2.28	0.25	Lapped at least 100 mm	Welded or welted	Not under pressure met in normal construction	Highly resistant to corrosion. If soluble salts are present, protect as for lead	May strain masonry. Not easy to work on site, so not suitable cavity trays
Bitumen							
Hessian base (class A of BS 6398)	3.8	—	Lapped at least 100 mm	Lapped at least 100 mm and sealed	Likely to extrude under heat and moderate pressure but this is unlikely to affect resistance to moisture penetration	The hessian or fibre may decay but this does not affect efficiency if the bitumen remains undisturbed. Classes D, E and F are most suitable for buildings that are intended to have a very long life or where there is risk of movement	Materials should be unrolled with care. In cold weather, warm before use. When used as a cavity tray, the d.p.c. should be fully supported. For further guidance see appendix B at BS 6398:1983.
Fibre base (class B of BS 6398)	3.3	—					
Asbestos base (class C of BS 6398)	3.8	—					
Hessian base and lead (class D of BS 6398)	4.4	—					
Fibre base and lead (class E of BS 6398)	4.4	—					
Asbestos base and lead (class F of BS 6398)	4.9	—					
High bond strength asbestos base	2.2	—					Should follow the recommendations in appendix C of BS 6398:1983
Polyethylene, low density (0.915 g/L to 0.925 g/L) complying with BS 6515	Approx. 0.5	0.46	Lapped for distance at least equal to width of d.p.c.	Welted	Not under pressure met in normal construction	No evidence of deterioration in contact with other building materials	Accommodates considerable lateral movement. When used as a cavity tray, may be difficult to hold in place and may need bedding in mastic for the full thickness of the outer leaf, to prevent rain penetration. Not suitable for use where compression stress is low, e.g. under copings.

Bitumen polymer and pitch polymer	Approx. 1.5	1.10	Lapped at least 100 mm	Lapped at least 100 mm and sealed	Not under pressure met in normal construction	Unlikely to be impaired by any movements normally occurring up to the point of failure of the wall	Accommodates considerable lateral movement. When used as a cavity tray, preformed cloaks should be used, e.g. at changes of level and junctions
B Semi-rigid							
Mastic asphalt complying with BS 1097 or BS 6577 of hardness appropriate to conditions	—	12	No joint problems	No joint problems	Liable to extrude under pressures above 0.65 N/mm^2	No deterioration	To provide key for mortar below next course of brickwork, up to 35% grit should be beaten into asphalt immediately after application and left proud of surface. Alternatively the surface should be scored whilst still warm
C Rigid							
Epoxy resin/sand	—	6.0	No joint problems	No joint problems	Not extruded	No evidence of deterioration in contact with other materials	Resin content should be about 15%. The appropriate hardener should be used
D.p.c. brick complying with BS 3921	—	Two courses, laid to break joint, bedded in 1:3 Portland cement:sand	No joint problems	Not suitable	—	No deterioration	Particularly appropriate where d.p.c. is required to transmit tension, e.g. in freestanding walls
Slate complying with BS 743	—	Two courses, laid to break joint, bedded in 1:3 Portland cement:sand	No joint problems	Not suitable	—	No deterioration	—

Source: Table 12 BS 5628 Part 3:1985

Table 4.10 Relevant British Standards

Material	Relevant British Standard
Bitumen	BS 6398
Brick	BS 3291
Copper	BS 2870
Lead	BS 1178
Mastic asphalt	BS 1097 and BS 6577
Slate	BS 743
Polyethylene	BS 6515
All others	BS 743

4.3 Clay bricks

These bricks are one of the most efficient materials to use in terms of their energy consumption. The brick factory is sited where extraction for clay is carried out and waste materials can reprocessed immediately. Badly fired bricks can be used to block up the openings to kilns, or sold as seconds for hardcore.

Firing of clay bricks

The act of firing or subjection of minerals to heat has various effects which are due not just to a straightforward drying out of material. The first effect is a sintering of particles at low temperatures which bond together through a diffusion process. This still leaves a degree of porosity between the particles which will affect the strength of the material, its absorbency and overall durability. These voids can be reduced with applied pressure. This bonding mechanism is used in the manufacture of concrete blocks. Higher temperatures, which in brickwork will be over 1000°C, will have a more radical affect; there will be chemical change and the feldspar and silica minerals will melt and flow around the clay particles to form a glassy matrix, changing their composition to new silicates and oxides. There will be less porosity and greater strength.

Although the technology of making ceramics is based on tradition and experimentation, and the phase changes are complex, they can be predicted to some extent from phase diagrams. These relate the proportion of minerals used to the temperature of the mixture, predicting phase change and the final structure of the material. The main mineral constituents will be flint (SiO_2), feldpsar (leucite) and clay (mullite). Chemical changes are also accompanied by subtle changes in the configuration of silicates which change their geometry, rotating their bonds and forming tighter packing arrangements, producing denser materials with a decrease in their overall size.

As a true ceramic, brickwork is subject to brittle behaviour under load with no modulus of elasticity. However, the material is not static and even after firing it will be responsive to the effects of heat and moisture showing incremental changes in length. The degree of porosity will affect durability in terms of resistance to frost attack.

Bricks are traditional ceramics, made from clays (alumina

Figure 4.3 Prairie house by Frank Lloyd Wright. Arthur Heurtley House, Oak Park, Illinois, USA, 1902. The horizontal banding is from the slight protrusion of brickwork courses.

and silica) which are tamped into moulds and then fired after air hardening, sometimes called *leather hardening*. The variety of brick colours comes from the regional variation in clay deposits where there is a change in mineral content, in combination with the temperature of firing and amount of oxygen available. As iron oxide content increases, bricks change from white (which have under 6% iron oxide) to light yellow, to orange and then red. With iron oxide concentrations between 8 and 10% in the clay deposits as found in the coal measures of Staffordshire, bricks can turn blue or purple as red iron oxide is converted into a black oxide, especially under conditions of oxygen reduction in firing.

Main clay deposits for bricks

There are nine major geological variants in brick clay and shales classified by age in million of years which also reflects their history of weathering and deposition and broadly the types of minerals present. All clays are classified as such by their particle size, which is less than two microns in diameter. Erosion of material to this size results in a particular grouping of minerals with certain properties. Their exact structure was unknown until examination by X-ray and electron microscope techniques only available in the mid-twentieth century. There are three main minerals recognized as important components in the clay used for brickmaking which give the right characteristics for working the clay into bricks and firing to make a predictable and durable building material. These are kaolinite, produced in non-marine environments under tropical conditions and illite and montmorillonite from marine environments. All clay deposits consist of hydrated aluminosilicates with a predominance of silica and alumina but with varying amounts of potash (K_2O), lime (CaO), soda (Na_2O), magnesium (MgO) and iron (FeO, Fe_2O_3). There may be some carbon from the breakdown of organic material.

Paleozoic

Devonian (395–345 million years B.C.) Exeter, South Wales. Hard shales giving fireclay.

Carboniferous (345–280 million year B.C.) Scotland, North England, Midlands. Grey shale and clays in coal measures from sub-tropical swamps (illite and kaolinite) producing hard dark red bricks. Etruria Marl results from the weathered material from coal measures and produces blue bricks and tiles.

Mesozoic

Triassic (215–190 million years B.C.) Nottinghamshire, Leicestershire Keuper Marl, red-brown mudstone from desert environments with deposits from saline inland lakes and seas (illite) producing light red and cream bricks.

Jurassic (190–136 million years B.C.) These beds were created from marine and estuarine environments with advancing and retreating sea levels (illite and montmorillonite) and from Lincoln to Cheltenham, Oxford clays produce light pink flettons with a high carbon content.

Cretaceous (136–65 million years B.C.) In Sussex and Dorset the Wealden clays and in Kent, Dorset, Wiltshire and Cambridgeshire the Gault clays were deposited from the same mechanisms operating in the Jurassic period.

Tertiary

Eocene (54–38 million years B.C.) The Berkshire, Suffolk, Dorset purple and grey clays known as the Reading Beds and the London clays running from Suffolk to London and Hampshire were a result of river deposition into inland basins and are generally known as ball clays from the shape they formed in extraction and transportation (kaolinite). Sand was often added to prevent warping in these clays.

Quaternary The pleistocene period produced clays as a result of the ice age from glacial deposits with a mixed mineral profile.

Lower Pleistocene (2.5–0.5 million years B.C.) Red bricks are produced from Icenian clays in East Suffolk.

Middle and Upper Pleistocene (0.5–0.1 million years B.C.) North England, Wales and East Suffolk has Glacial boulder clays with flint-rich clays on the Hertfordshire/Buckinghamshire border.

Holocene (0.1 million years B.C. to present day) There are alluvial clays from river deposition in the York, Humberside, Lincolnshire, East Anglia, Thames and Medway valleys.

Kaolinite This is a hydrated aluminosilicate and is the most common mineral with the simplest chemical structure as weathering can reduce it no further. Under pressure in deep sediments and at high temperatures (over 100°C) reactions can transform the mineral back into illite and montmorillonite. Firing transforms the mineral, producing needle-like crystals of mullite in a silica matrix. This mineral is also used to produce pure alumina after a reaction with hydrogen sulphide to yield high-quality engineering ceramics. It may be referred to as china clay from the direct decomposition of feldspar crystals in granite and is often used as filler material for paper making or paint production.

Illite This potassium aluminium aluminosilicate is part of the mica group of sheet silicates with a monoclinic structure. Potassium cations are held between the sheet silica structures.

Montmorillonite This hydrated aluminium silicate with an amorphous structure expands and contracts greatly under varying moisture conditions. This mineral has calcium and sodium cations as well as water molecules held between the sheet silicate structures and these attract the greater numbers of water molecules in saturated conditions to give the characteristic swelling.

The firing of these minerals does not replicate their original state before they weathered. Natural mineral structures in rocks have usually taken time to form and slow cooling produces crystalline structures. Industrial firing techniques will produce glassy silicates which bond around sand and other particles to form a solid unit. In ceramic reactions generally under the effects of heat, the original minerals do not totally decompose but change their three-dimensional configuration, called *topotaxy*.

Plasticity This is a term used to describe clay cohesion when clay is mixed with water and this property will vary with different clay minerals. As a property plasticity is dependent on the mineralogy of clay deposits and their electrochemistry which determines the amount of water that can be bonded locally and which allows particles to slip over each other with ease. Water is held within the clay body by electrostatic attraction and only starts to be driven off at 105–110°C; the clay then ceases to be plastic and is unworkable, a state that can be reversed by adding water. The remaining water content is part of the clay chemistry and will only start to be removed at about 450°C. At this temperature and higher the reaction is irreversible. Amorphous products are formed which then transform into clearer structures at temperatures over 1100°C. There is a plasticity index for clays which allows for predicting plastic behaviour. This relates the water content needed for basic workability known as the *plastic limit* with the water content at which the clay will begin to flow known as the *liquid limit*.

Making bricks

Most clays have to be prepared by being sieved to remove stones, ground to achieve the right particle size for sintering and modified with the addition of water for the correct moisture content for processing. Early production methods used to leave excavated clay to weather allowing the action of rain and frost to break up the clay into finer particles, ready for kneading.

Bricks are made in three basic ways:

(1) Pressed into moulds to give individual units such as Flettons.
(2) Extruded and then wire-cut to give solid or perforated bricks, providing the paramount method for making bricks other than flettons and accounting for the means of production of one third of all facing bricks.
(3) Hand-made from soft clay which is rolled often in sand and sawdust and pressed manually into moulds.

This order of making also reflects the overall quality of finish produced. The pressed bricks have the smoothest finish, the extruded bricks have a greater texture to them and the hand-made bricks are uneven and have individual character. The quality of colour is affected by textural difference. The coarser the brick, the greater the scattering of light waves and the softer the colour.

Most bricks are moulded to give a frog on one side, which should not exceed 20% of the volume of the brick. This makes the brick lighter and faster to lay, although the frog should be laid uppermost and then be fully filled with mortar to optimize the strength of the unit. For walls that are not taking direct load and for most domestic house construction, the high strength of the bricks used means that placing a brick frog down, unfilled with mortar, may be acceptable.

Brick types

The range of bricks available is dependent on the origin of the clay and the method of manufacture. It has always been a significant industry in Britain and, although reduced in size from the nineteenth century, there are still 55 manufacturers and many of these have several works. For a comprehensive listing of brick manufacturers and their products see *The Architects Journal* supplement 27.11.85, and the current edition of *Specification* also published by the Architectural Press.

Rubbers Rubbers are bricks that are made from well-washed red clay which removes all coarse particles with a high proportion of sand. The bricks are low-fired just short of vitrification, the particles are not sintered together and although they can be *rubbed* to shape they are also subject to erosion. They are consequently soft, can be easily cut into a rough shape, and then rubbed to an exact shape, e.g. for voussoirs for arches. These were not advocated for external use although they are common, especially in London in combination with yellow stocks.

Commons Commons are bricks that have a minimium amount of processing before being fired. They are likely to be made straight from clay that has not been washed.

Originally there were three classes of bricks which included *malms, washed* and then *commons*, all categorized by the way that earth was prepared. Malm referred to a process which added chalk and screened the mixture, washed did not screen the material in the same way and commons involved the minimum amount of preparation. Today the term is used to describe bricks that are pink or red in colour.

Flettons Flettons originate from Fletton near Peterborough and are made by the London Brick Company. They are cream to pale pink and are made from clay that needs little preparation and so are also referred to as *commons*. They often have sand-blasted finishes as facing bricks. Today the term Fletton is used in a more general way to describe bricks made from *Oxford clay* which has a high carbon content (allowing for ease in firing), also a moisture content which allows for very little preparation of the clay. With applied finishes and the use of pigments, the London Brick Company now has a range of 15 colours, so clay origin is no longer a limiting factor on the final colour that can be achieved. This kind of variety can be made by many brick manufacturers, but it does remove the specificity of place, where local architecture is immediately recognizable with a regional identity relating to local materials.

Stocks Stocks refer to bricks that have been hard burnt and have some variation in colour. Hard stocks are over burnt and although sound, can be too discoloured for ordinary work. They were used traditionally for pavings, footings and in the body of thick walls.

Engineering bricks Engineering bricks are generally high strength and used where durability is required and/or where great resistance to water penetration is needed, such as in inspection chambers. There are the traditional colours of blues, purples and dark reds although the classification of density and strength can be met by ordinary *commons* which will be lighter-coloured reds.

4.4 Specification of bricks

BS 3291: 1985 is the British Standard specification for clay bricks. The parameters for specification are by size, durability, efflorescence, compressive strength, and water absorption. The rate of suction may be relevant with calculated masonry, and pre-wetting of brickwork may need control if specified mortar strengths are to be reached. Water is needed in sufficient quantities for the chemical reaction with cement and if this is diminished in any way by over-absorption by the bricks, there will be subsequent loss of strength to the mortar. Conversely, if there is inadequate absorption by the brickwork there will be poor adhesion of mortar in the first instance and unsatisfactory bonding.

Workmanship

Workmanship is an important factor and can affect the calculated strength of brickwork by 30% or more. The characteristic compressive strength of a masonry wall never achieves the compressive strength of an individual unit (whether brick or block), and at best would only be half the strength for low compressive strengths of masonry and can be only one fifth; this can be further reduced by poor supervision on site and poor workmanship. See Table 4.11 which shows how the compressive strength of masonry will normally be less than the predicted strength of the masonry unit.

Comparable strengths

Bricks vary greatly in their strength and range from $7\ \mathrm{N\ mm^2}$ to over $100\ \mathrm{N\ mm^2}$ in their compressive strength. The compressive strength of a brick is evaluated from an average of test on ten bricks sampled and tested in accordance with BS 3291, with frogs filled. As strength is related to density, there is a correlation between strength and porosity, with water absorption as an indicator. (See BS 5628 CP for use of masonry; Part 1 for details of this relationship to flexural strength.) Water absorption is controlled for engineering bricks and DPC classes of bricks. The new CEN standard will make density part of the brick specification for the first time in the UK. (See Table 4.12.)

However, as the pore system of a brick is not continuous, water absorption is restricted to the outer shell and complete saturation may not occur. Consequently water absorption is used as a parameter for general durability and is more relevant to predictions with regard to frost resistance.

The actual compressive strengths needed for non-load bearing partitions and domestic building are fairly low and it is acceptable in these situations to use bricks laid frog down to speed the process of bricklaying. However, in order to meet a higher compressive strength for structural brickwork carrying floor slabs the frogs should be laid upwards as previously stated and completely filled with mortar.

Reinforced brickwork

The strength of masonry generally can be improved by reinforcement (see *MBS: Structure and Fabric Part 2*). BS 5628 Part 2 deals with reinforced masonry by specifying the structural units, and accompanying aggregates, cements

Table 4.11 Compressive strengths of masonry walls

Construction: Constructed with standard format bricks — Compressive strength of unit (N/mm^2)

Mortar designation	5	10	15	20	27.5	35	50	70	100
i	2.5	4.4	6.0	7.4	9.2	11.4	15.0	19.2	24.0
ii	2.5	4.2	5.3	6.4	7.9	9.4	12.2	15.1	18.2
iii	2.5	4.1	5.0	5.8	7.1	8.5	10.6	13.4	15.5
iv	2.2	3.5	4.4	5.2	6.2	7.3	9.0	10.8	12.7

Construction: Constructed with blocks having a ratio of height to least horizontal dimension of 0.6 — Compressive strength of unit (N/mm^2)

Mortar designation	2.8	3.5	5.0	7.0	10	15	20	35 or greater
i	1.4	1.7	2.5	3.4	4.4	6.0	7.4	11.4
ii	1.4	1.7	2.5	3.2	4.2	5.3	6.4	9.4
iii	1.4	1.7	2.5	3.2	4.1	5.0	5.8	8.5
iv	1.4	1.7	2.2	2.8	3.5	4.4	5.2	7.3

Construction: Constructed with hollow blocks having a ratio of height to least horizontal dimension of between 2.0 and 4.0 — Compressive strength of unit (N/mm^2)

Mortar designation	2.8	3.5	5.0	7.0	10	15	20	35 or greater
i	2.8	3.5	5.0	5.7	6.1	6.8	7.5	11.4
ii	2.8	3.5	5.0	5.5	5.7	6.1	6.5	9.4
iii	2.8	3.5	5.0	5.4	5.5	5.7	5.9	8.5
iv	2.8	3.5	4.4	4.8	4.9	5.1	5.3	7.3

Construction: Constructed from solid concrete blocks having a ratio of height to least horizontal dimension of between 2.0 and 4.0 — Compressive strength of unit (N/mm^2)

Mortar designation	2.8	3.5	5.0	7.0	10	15	20	35 or greater
i	2.8	3.5	5.0	6.8	8.8	12.0	14.8	22.8
ii	2.8	3.5	5.0	6.4	8.4	10.6	12.8	18.8
iii	2.8	3.5	5.0	6.4	8.4	10.0	11.6	27.0
iv	2.8	3.5	4.4	5.6	7.0	8.8	10.4	14.6

Source: Table 2 BS 5628 Part 1:1992

Table 4.12 Compressive strengths and water absorption

Class	Compressive strength (N/mm^2)	Water absorption (% by mass)
Engineering A	≥70	≤4.5
Engineering B	≥50	≤7.0
Damp-proof course 1	≥ 5	≤4.5
Damp-proof course 2	≥ 5	≤7.0
All others	≥ 5	No limits

Note: There is no direct relationship between compressive strength and water absorption as given in this table and durability. Damp-proof course 1 bricks are recommended for use in buildings whilst damp-proof course 2 bricks are recommended for use in external works.

Source: Table 4 BS 3921:1985

Table 4.13 Cover of concrete for carbon steel

Exposure situation	Thickness of cover (mm)			
E1	20	20	20	20
E2	—	30	30	25
E3	—	40	35	30
E4	—	—	—	60
Concrete grade in BS 5328	25	30	35	40
Min. cement content (kg/m^3)	250	300	350	350

Source: Table 14 BS 5628 Part 2:1985

and mortars as in Part 1, and includes steel for reinforcement. Prestressing is also covered in this part of the code. If there is a net shrinkage in the masonry due to the specification of concrete or calcium silicate masonry structural units, there will be a loss of prestressing in tendons, but this is calculable and can be allowed for. This is not relevant to the expansion of clay-fired structural units and any moisture expansion of these components can be ignored. If reinforced, prestressed structures are contemplated, care must be taken at an early stage to design for the detailing of anchorage points. In reinforced masonry attention must be given to the protection of reinforcement. Table 4.13 gives details for carbon steel relating to the cover of concrete required, the grade of concrete cover specified and the degree of exposure. Table 4.14 provides specifications for approved coatings to reinforcement in the absence of satisfactory cover.

The exposure situations grade from:

- **E1** — for internal or external work which is classsified as sheltered or very sheltered (defined in BS 5628 Part 3: 1985).
- **E2** — for buried masonry or masonry submerged in fresh water or externally where the exposure category is sheltered/ moderate or moderate/severe.
- **E3** — for masonry subject to wet and freezing situations, cyclical wetting or externally where the exposure category is severe or very severe.
- **E4** — for masonry exposed to corrosive situations whether through salted or acidic waters.

If reinforcement is placed in bed joints the mortar cover from the exposed face of the masonry should be a minimum of 15 mm. Movement joints should still be provided as outlined in BS 5628: Part 3:1985 clause 20. If joints have

Table 4.14 Reinforcement coatings

Exposure situation	Minimum level of protection for reinforcement, excluding cover	
	Located in bed joints or special clay units	Located in grouted cavity or Quetta bond construction
E1 Sheltered	Carbon steel galvanized following the procedure given in BS 729. Minimum mass of zinc coating 940 g/m^2	Carbon steel
E2 Moderate to severe exposure to fresh water	Carbon steel galvanized following the procedure given in BS 729. Minimum mass of zinc coating 940 g/m^2	Carbon steel or, where mortar is used to fill the voids, carbon steel galvanized following the procedure given in BS 729 to give a minimum mass of zinc coating of 940 g/m^2
E3 Severe exposure	Austenitic stainless steel or carbon steel coating with at least 1 mm of stainless steel	Carbon steel galvanized following the procedure given in BS 729. Minimum mass of zinc coating 940 g/m^2
E4 Exposure to salts and water	Austenitic stainless steel or carbon steel coated with at least 1 mm of stainless steel	Austenitic stainless steel or carbon steel coated with at least 1 mm of stainless steel

Source: Table 13 BS 5628 Part 2:1985

been designed for contraction movement only in concrete masonry units, these should be supplemented at 30 m intervals by expansion joints.

Sound insulation

To optimize the sound insulation of walls, and to meet the Building Regulations requirement for separating walls, bricks should be laid frog up. Quality control is also important here. Most party walls although built in 215 mm brickwork, do not comply with the Building Regulations in practice due to poor workmanship, inappropriate details and lack of thought in providing junctions which are discontinuous so that impact sound is not transmitted. As an example of performance a half brick wall plastered both sides should give 45 dB reduction, and a one-brick wall plastered both sides should achieve a reduction of 50 dB.

Thermal performance

With regard to the thermal performance of walling systems, even the lowest density bricks have such a poor performance thermally that the outer leaf of a wall may only contribute about 5% to the overall thermal resistance (Table 4.15). In order to achieve a U value of 0.45 W/m^{2}K Table 4.16 can be used to establish the thickness of insulation required, dependent on what type of insulation is chosen and inner leaf material used for cavity wall construction.

Brickwork mortars

It is artificial to consider the strength of a brick unit in isolation. Mortar is fundamentally a matrix for brickwork acting as a bonding mechanism as well as a medium for taking up dimensional variations. Table 4.11 gives the characteristic compressive strength of masonry allowing for different mortar variables.

The mortar designations referred to are given in Table 4.5. The weaker the mix, the more able the wall to accommodate movement as a result of temperature. The weaker mixes are also able to resist frost attack more readily but are less resistant to water penetration and moisture, at the expense of overall strength.

The components of mortar should also be separately specified. If mortar is ready mixed it should comply with BS 4721. Problems with mortars, including efflorescence and staining can be due to contamination and poor constituents of mortar. They should always be specified properly.

Cement

Portland cement to BS 12 (ordinary and rapid hardening)
Portland blast-furnace cement BS 146: Part 2
Sulphate-resisting Portland cement BS 4027
Masonry cement BS 5224

Lime To BS 890.

Table 4.15 Properties of cavity insulation materials

Cavity type	Product	British Standard	Density (kg/m^3)	Thermal conductivity (W/m K)
Partially filled	*Rigid slabs or boards*			
	Expanded polystyrene bead board	3837	15–20	0.034–0.037
	Extruded expanded polystyrene		20	0.028
	Mineral fibre		23–50	0.033–0.037
	Polyurethane boards		32	0.022
	Polyisocyanurate boards		28	0.022
	Foamed glass slabs		125	0.042
	Glass fibre slabs		16–28	0.033–0.037
Fully filled	*Batts*			
	Mineral/glass fibre	6676	16–50	0.033–0.037
	Blown			
	Mineral wool or fibre	6232	30–70	0.037–0.040
	Glass fibre		18–22	0.031
	Polystyrene beads/granules		12–18	0.035–0.040
	Foamed			
	Urea formaldehyde	5617	10	0.040
	Polyurethane foam		11–16	0.037
	Polyurethane foam (cavity stabilization)		36–40	0.023

Source: Table 2 *Improved standards of insulation in cavity walls with outer leaf of facing brickwork* March 1990 Brick Development Association

Table 4.16 Insulation thickness criteria

Inner leaf			Internal surface finish	Thermal conductivity of cavity insulation (W/m K)	Thickness of insulation required for U-value of 0.45 W/m^2 K (mm)							
					Partially filled cavity				Completely filled cavity			
					Inner leaf block thickness (mm)				Inner leaf block thickness (mm)			
Type	Density (kg/m^3)	Thermal conductivity (W/m K)			100	125	150	200	100	125	150	200
Ultra light-weight block	475	0.11	Wet plaster	0.022	17	12			21	16		
				0.035	27	19	11	0	33	25	17	1
				0.040	31	22	12	0	38	29	20	1
				0.045	34	24	14	0	43	32	22	1
			Plasterboard drylining	0.022	13	8			17	14		
				0.035	21	13	5	0	27	22	12	0
				0.040	24	15	6	0	31	26	13	0
				0.045	27	17	7	0	35	29	15	0
Light-weight block	600	0.19	Wet plaster	0.022	25	22			29	26		
				0.035	40	36	31	22	46	42	37	28
				0.040	46	41	35	25	53	48	42	32
				0.045	52	46	40	28	59	54	48	36
			Plasterboard drylining	0.022	22	19			25	22		
				0.035	35	30	25	16	41	36	32	22
				0.040	40	34	29	32	46	41	36	26
				0.045	44	38	32	21	52	46	41	29
Medium density block	1400	0.51	Wet plaster	0.022	32	31			36	35		
				0.035	52	50	48	45	58	56	55	51
				0.040	59	57	55	51	66	64	62	58
				0.045	66	64	62	58	74	72	70	66
			Plasterboard drylining	0.022	29	28			33	32		
				0.035	46	54	43	39	52	51	49	46
				0.040	52	51	49	45	60	58	56	52
				0.045	59	57	55	50	67	72	63	59
Dense block	2300	1.63	Wet plaster	0.022	35	35			39	39		
				0.035	56	56	55	54	63	62	62	61
				0.040	65	64	63	62	72	71	70	69
				0.045	73	72	71	70	81	80	79	78
			Plasterboard drylining	0.022	32	32			37	35		
				0.035	50	50	50	49	57	56	56	55
				0.040	58	58	57	56	65	64	64	63
				0.045	65	65	64	63	73	72	72	71
Brick-work	1800		Brickwork	0.022	35				43			
				0.035	56				68			
				0.040	64				77			
				0.045	73				87			

Source: Tables 3 and 4 *Improved standards of insulation in cavity walls with outer leaf of facing brickwork* March 1990 Brick Development Association

Aggregates

Natural aggregates to BS 882 or BS 1200

Lightweight aggregates to BS 877 Part 2 or BS 3797 Part 2

Pigments To BS 1014.

White clay bricks can be whitened further by adding chalk, and cream colours can be produced if lime is added to bricks with some iron oxide. As lime and chalk are both calcium carbonate the amounts used should be limited to below 15% as above this figure the amount of carbon dioxide liberated in firing will make the bricks too porous.

Greater amounts of iron oxide will give brown colours, manganese dioxide will turn red clays dark brown and buff clays grey.

Plasticizers To comply with BS 4887, but only with the designer's written permission.

Plasticizers should not be used unless permission is given by the designer, and calcium chloride or frost inhibitors based on this compound should not be used.

Water To BS 3148, pure enough for drinking.

Brick colour

Bricks can undergo a further firing with an applied glaze. These bricks are second-fired at higher temperatures and are of better quality. Oxides will give different colours, for example, copper oxides will give green, cobalt oxides give blue, and iron oxides red. If the oxygen content is reduced in firing then oxygen will be taken from the oxides to combine with vapourized carbon and the colours change. The copper oxide will then produce purples. Pigments can be added to glazes to produce an even greater variety of colours, and in this situation, iron oxides will yield black.

Brick textures

Bricks are usually pressed into moulds profiled for frogs. The frogs should not have a depression that exceeds 20% of the volume of the brick. The moulds can be lined with sand which will give a sharper finish and different face characteristics. Not only sand but also different iron oxides and ceramic pigments can be added to the face before firing. These are commonly called *facing bricks* and are vulnerable to being chipped. Colours can be changed as described instead of altering the whole body colour. Bricks can be cut from extruded slabs of clay without frogs, and these are referred to as *wirecut*. Wirecut bricks are often perforated to achieve lightness and also provide an opportunity for placing reinforcement. The amount of perforation should not exceed 25% of the gross volume of the brick and the area of any hole should not exceed 10% of the gross area of the brick. The solid area measured across the width of the brick should not be reduced to below 30% of the brick.

Special shaped bricks

There is a range of *standard specials* which are outlined in BS 4729: 1971. The advantage of specifying standard specials is that they are likely to be in stock. Special shapes can be purpose-made but they are expensive and can have long delivery times. There used to be a greater range of brick shapes which could be built up with fine joints to form string courses or full cornices, and some companies used to make a full cornice detail. Often the old moulds can still be found for restoration work. Drain bricks were also a standard traditional pattern for forming channels in buildings, particularly agricultural sheds.

Figure 4.4 Ark Building, Hammersmith, London 1991, Ralph Erskine. Instead of using special bricks, these ordinary bricks for keying plaster have been laid imaginatively and courses of ordinary bricks bush hammered to achieve a rich texture.

There is a range of proprietary bricks that are commonly specified. They include the *Phorpres* keyed brick which has integral dovetailed keys for plastered or rendered finishes. The Phorpres cellular brick is 75% of the weight of a fletton and has three cells.

There are some large format bricks known as *Calculon*, made by the London Brick Company and Redland Bricks Ltd, used for internal load bearing partitions of 175 mm thickness. They have special hand holds and are lightened by incorporating holes for ease of laying. Manufacturers show a 30% increase in productivity and a 40% saving in mortar in comparison with a full brick wall. There are optional types, either smooth-sided or keyed for plaster finishes.

Brick sizing

Sizes have varied due to locality, generally being larger in the North of England and Scotland. Original sizes are imperial and vary from $8\frac{3}{4}''$ in length, $2\frac{1}{2}''$ thick and $4\frac{1}{4}''$ wide in London, to $2\frac{7}{8}''$ in height in the Midlands and the North, to 2″ high in the South. Rivington (1879) lists 19 different standard sizes for bricks. Sizing is historical in origin, from the Roman use of shallower bricks. The underlying ratio to allow for proper bonding, is that the length of a brick should be determined by twice the width, plus one mortar joint.

The British Standard now uses one co-ordinating size, 225 × 112.5 × 75 mm, which is slightly smaller than the original imperial equivalent, allowing for 10 mm on each dimension for mortar joints. The actual size, 215 × 102.5 × 65 mm, is the work size.

There is a modulating size of brickwork described in BS 6649:1985 as 200 × 100 × 75 for the co-ordinating size. There seems no additional financial advantage in using this brick although it makes the dimensional co-ordination possibly easier. See revised BS 4729.

There is currently work being done towards a European standard although sizes of bricks have not been discussed in detail. European practice tends to use clay blockwork, rendered with 25mm or more of render. Northern European countries tend to use vertically perforated blocks and Mediterranean countries use horizontally perforated blocks. Our smaller UK brick unit is in evidence only in Holland and Denmark.

It is important in using bricks to ensure that building dimensions are calculated as multiples of the brick and mortar unit. The work size is the actual size of a brick within permissible limits and the co-ordinating size allows for joints. The BDA publish tables in their Design Note 3, *Brickwork dimensions*, and advocate notation on working drawings showing the number of bricks to be used as well as the co-ordinating dimension. Traditionally the overall dimensions of 24 brick units must fall between specified limits in the UK. This is not standard practice in Europe, as bricks and blocks are measured as discrete units and results expressed as the mean of 100. Dimensional tolerances are worked out relative to the square root of the work size. It is likely that the UK system will be phased out as European systems take over.

Moisture expansion in brickwork

When bricks are fired they lose most of their water content. When exposed to the atmosphere after firing they have the ability to take up moisture which can cause an expansion of about 0.1%. They should have a period of two to three weeks to adjust on site before being used. There is a risk of cracking if they are delivered straight from being fired as only bricks fired at very high temperatures will lose all of their moisture and be totally stable. After firing, brickwork will expand as water vapour is absorbed, but the material stabilizes after 14 days and expansion afterwards is minor; ideally, new bricks should be laid after this period has elapsed. Clay brickwork expands irreversibly by 0.02–0.07%; this coupled with a reversible movement of ±0.02% would give a maximum movement of 0.09%. Accordingly it is normal to allow for an expansion rate of 1 mm for every metre run of brickwork. There is a levelling out of this effect and no more water will be absorbed approximately three months after the bricks were first made. There is a greater adjustment in respect of take up of moisture from bricks which have been fired to a more vitrified state. This is understandable as the moisture loss in firing would be complete and the combination of atmospheric pressures and relative humidity would force a take up of moisture.

Moisture movement is also related to the types of clays used, as shown in Table 4.17.

4.5 Deterioration of brickwork

The ultimate deterioration of brickwork is related to the density and porosity of the brick as well as the quality of the clays used in production, and final detailing. As all the decaying mechanisms of masonry are related to water and its effects, density and porosity are important parameters which determine the amount of saturation that can take place. Bricks, with a compressive strength of about 20 N/mm^2, are generally regarded as durable for most conditions. For more exposed situations, a strength above 30 N/mm^2 would be more appropriate. Deterioration is usually dependent on water saturation, whether through the action of freezing or attack by sulphate reactions with mortars, or by the crystallization of different salts after subsequent water loss. See Table 4.12.

There are identifiable areas which are more exposed in a building. The most severely exposed areas are on free-standing external walls, inspection chambers, brickwork below damp-proof course level, parapets and external walls.

Table 4.17 Brickwork expansion rates

Clay type	Expansion in one-metre run (mm)
Carboniferous	1.0
Lower Oxford, Keuper Marl	0.5
London stock gault	0.25

Sulphate attack

There is one compound in cement, tricalcium aluminate, which can react with sulphates if water is present, forming calcium sulpho-aluminate (often referred to as ettringite). This has a greater volume than the original compound, and in expanding after the chemical reaction will deform brickwork. This new compound is softer and, being less dense, is ineffective as a mortar and will ultimately crumble and decay. The reaction is more extreme with magnesium or sodium sulphates and will occur if the bricks are persistently damp. These reactions are more likely to happen if a mortar mix is more porous, for example, a 1:1:6 mix which will enable water to be more easily retained.

Knowledge of brick performance is derived from experience with different brick types. As well as choosing the ideal brick for performance in a particular situation, the detailing at eaves, gables and door and window openings should protect the brickwork. An appropriate mortar mix should be chosen with sulphate-resisting cement if appropriate, and sand in mortars should also be carefully specified to avoid contaminants.

Freezing mechanisms in brickwork

See *MBS: Finishes* Chapter 3 (Frost and freezing mechanisms).

There are three basic categories of bricks:

F — Frost resistant These bricks should be used externally in situations of high exposure to repeated freezing and thawing. Cyclic testing which imitates the freeze and thaw cycles can give an indicator of the likely performance of these bricks which should survive 100 cycles of testing in a saturated condition.

M — Moderately frost resistant These bricks will still be appropriate for external walls in a building but unsuitable for more exposed conditions such as garden walling.

O — Not frost resistant These bricks should only be used internally and can be damaged if delivered on site and exposed in winter conditions.

As a general rule all bricklaying should be stopped when the temperature on site falls below 3°C. Additives should not be used to reduce the freezing point of water as they can affect the hydration and strength of cement. Air entraining can be a better solution giving larger air voids for the expansion of ice during freezing conditions. Unless the air content can be guaranteed at 18% this method should not be used, as the strength of the mortar will be adversely affected. Agreement is needed for this in case it affects the strength of the mortar and falls outside that specified.

Bricks do not need pre-wetting in winter conditions as the rate of evaporation is slower. Greater protection is needed from the elements, and brickwork should be well covered with polythene laid over the top of hessian or another insulant to prevent saturation in wet conditions. (See *Brick laying in winter conditions*, BDA Note 3, January 1986.)

Efflorescence

Efflorescence arises when soluble salts lose water and crystallize. The same salts of magnesium and sodium sulphate are involved as in direct sulphate attack, but their effect is not as disastrous and does not cause a structural weakening of the whole walling system. Often these salts crystallize on the surface of brickwork and, being white (and fluffy in the case of magnesium sulphate), are unsightly. However, if they crystallize in the pore walls just below the surface skin of brickwork, this is known as *subflorescence*. As the salts crystallize they expand in volume and can act like a wedge driving the outside skin off the face of the brickwork. This can be damaging and disfiguring to the facing fabric of the building. Other salts include potassium and sodium sulphate but they form glassy skins which are not as noticeable. Sometimes efflorescence occurs when new bricks containing sulphates are saturated in the building process. The resulting salts will then form until the wall has lost all of its water. This could happen over a period of 18 months to two years. If salts are still left in the mortar of the brickwork and there are extreme driving rain conditions which will constantly saturate the wall, the efflorescence can continue for far longer. If salts are being sucked up into the brickwork from ground waters, due to the lack of an adequate damp-proof course, then deterioration will be progressive, eventually forming a distinctive *tidemark* showing the extent to which capillary action can occur (normally about 600 mm).

White efflorescence, which is more prominent in spring, can be cleaned by brushing away and the deposits should be carefully collected so as not to contaminate lower levels of brickwork. The remainder should be removed with clean cold water applied with a damp sponge. Other yellow or green efflorescence caused by the crystallization of vanadium salts should be allowed to disappear after weathering. As a last resort there are chemical treatments, details of which can be found in BDA Building Note 2, September 1986.

Heavy efflorescence is defined in BS 3921 as affecting over 50% of the walled area and/or powdering or flaking of the surface, with moderate efflorescence between 10 and 50% and slight efflorescence being evaluated at less than 10%. However, even the category described as slight efflorescence would be a noticeable defect to both architect and client.

Figure 4.5 Cascades, Docklands (1989). This raking apartment building by Campbell Zoglovitch Wilkinson and Gough also suffered with efflorescence after completion.

European Standards

The new European Standard is working towards three classifications which relate to the amount of saturation and the following is quoted from *European clay standard takes shape*, Building Technical File, No 29, April 1990, published by the BDA.

- Brickwork in which the clay units are saturated with water, or very nearly.
- Brickwork in which the clay units may be damp but which is protected from near saturation by the design detailing of the building.
- Brickwork in which the clay units are kept completely dry (for example, by an impermeable cladding) or which are never exposed to freezing temperatures (such as when used internally).

With regard to soluble salt content, there are two categories:

- Low (L) which has upper limits for the following salts:
  - Calcium 0.003%
  - Magnesium 0.030%
  - Potassium 0.030%
  - Sodium 0.030%
  - Sulphate 0.500%
- Normal (N) which has no limits.

This categorization has been adopted in the CEN draft, except the salts have been grouped as (sodium + potassium + magnesium) ions as one percentage value, and water soluble sulphate is grouped as another percentage value. Table 4.18 summarizes the approach.

In practice there are variations from expected performance where bricks with high salt content (i.e. up to 3%) have not given problems although bricks with a salt content of under 1% have had serious problems.

Mortars should be specified accordingly and sulphate-resisting cement used for situations where the bricks may be in contact with groundwater, or where there may be sulphates from flues.

In combination with frost resistance there will immediately be six categories for durability.

Brick Durability and Prevention of Decay

As most brickwork problems stem from moisture entrapment, it is tempting to try and prescribe proprietary products that act as water repellants which are often silicone based. Moisture should be allowed to move in and out of brickwork, and sealers preventing this action can entrap moisture which could lead to frost attack. If the application of these treatments is selective to particular areas, it can have unforeseen effects. For example, the treatment of

Table 4.18 Salt content

Category	Max. total by mass (%)	
	($Na^+ + K^+ + Mg^{2+}$)	SO_4^{2-}
L	0.09	0.5
N	0.25	1.6
U	No requirement	No requirement

Figure 4.6 This brickwork which on an exposed housing estate in Bristol has suffered continuously from efflorescence since being built some seven years previously. Constant saturation and drying by the wind show the repercussions of using materials that have too high a salt content.

parapets then increases the amount of water run off to lower surfaces as it cannot be absorbed by the upper levels of brickwork. This increased saturation at lower levels will lead to a build up of lichen. (See 'Water repellant surface treatment — a word of caution' from the *Brick Bulletin*, Spring 1988.)

To maximize the durable qualities of brickwork it cannot be looked at in isolation. The jointing of brickwork should be specified to prevent unnecessary holding of water and this means avoiding ironed-in recessed joints. Ideally a joint should be finished flush and then tooled with the edge of a trowel to give either a weather struck profile or with a special tool giving a concave section or *bucket handled joint*. Tooling compacts the mortar whereas raking out a joint leaves a more porous and friable mortar. Flush joints are better than raked joints but fine cracks can develop with shrinkage of the mortar and can encourage water penetration. (See 'Recessed joints', *Brick Bulletin*, Winter 1988.)

Often the decision made on the pointing methods used in brickwork are made on aesthetic grounds and yet some principles should be noted to optimize weathering performance and durability. It is advisable to avoid ironed-in recessed joints, as horizontal ledges can allow water and sulphur dioxide deposits to accumulate, leading to long-term deterioration through chemical change, frost and freezing mechanisms.

Movement in brickwork

Although bricks seem to be a stable material they are not. As the temperature changes, this may impose a range of 85°C (especially with the darker colours) in the United Kingdom. The consequence could be an increase in length of 0.07 mm per metre using a coefficient of thermal expansion of 8×10^{-6}/K. When loading conditions vary, brickwork may respond by elastic behaviour, i.e. extending under load and then returning to its original position, or it may creep permanently. When moisture is taken up or evaporated, brickwork responds by expansion and contraction. There is irreversible movement when brickwork is delivered from the kilns with total water loss. It then takes up moisture from the atmosphere quite naturally, causing expansion and reversible movement when there is cyclical wetting and drying. If this movement is restrained the stress incurred will cause cracking. Restraint in brickwork is largely due to the geometry of the construction. Vertical loading effectively pins the brickwork and stops horizontal movement, explaining why there is less restraint in parapets and garden walls.

Brickwork is very often used in conjunction with blockwork producing a composite construction. Concrete blocks and calcium silicate bricks are liable to shrinkage and the amount of movement is even greater; as there may be two different materials moving together with components changing at different rates. Walling should be built to cater for the very worst possible movement. Even greater problems are encountered with composite construction where brickwork is tied back to a concrete frame and where shrinkage in the concrete frame could impose loading and stress on brickwork.

There are natural points such as openings for windows and doors where brickwork can change dimensionally. However, there will be longer expanses of brickwork over these openings where movement joints will have to be incorporated. The design of movement joints is one of the most difficult design decisions to make and from an early stage brickwork designers should have these in mind. Fillers should be compressible and the BDA recommends cellular polythene, polyurethane or foam rubbers.

CP 121 recommends joints of 10 mm at 12 m centres for brickwork, 7.5–9 m for calcium silicate bricks and 6 m is the general requirement for concrete blocks. Walls with low restraint, which may be spandrel panels or parapets, will have approximately twice the amount of movement and

Table 4.19 Moisture movement and thermal properties of common building materials

Material	Reversible moisture movement	Irreversible moisture movement	Coefficient of thermal expansion ($\times 10^{-6}$/K)
Timber	±(0.5–2.5)*	—	4–70*
Steel (depending on type)	—	—	10–18 (depending on type)
Concrete	±(0.02–0.10)	−(0.03–0.08)	7–14 (depending on aggregate used)
Brickwork and blockwork			
Dense concrete aggregate concrete products	±(0.02–0.04)	−(0.02–0.06)	6–12
Lightweight aggregate concrete (autoclaved) products	±(0.03–0.06)	−(0.02–0.06)	8–12
Aerated (autoclaved) products	±(0.02–0.03)	−(0.05–0.09)	8
Calcium silicate brickwork	±(0.01–0.05)	−(0.01–0.04)	8–14
Clay brickwork	±(0.02)	+(0.02–0.07)	5–8

* Depending on measured direction.

Source: Table 1 *Designing for movement in brickwork* Design Note 10 Part 2 August 1988 Brick Development Association

joints ought to then be designed with centres of 5 m, or alternatively 20 mm expansion joints should be provided at 10 m centres. It is generally acknowledged that vertical movement is of the same order as horizontal movement. Movement joints will then inevitably divide the facade of the building up into bays.

See Table 4.19 for moisture and thermal properties of common materials.

4.6 Calcium silicate bricks

These are the results of attempts to make artificial stone in Germany in the nineteenth century from silica, sand, lime and water. The basic constituents have not changed although they may be described in terms of crushed flint or sand aggregates. The materials are slaked to ensure that the lime (between 5 and 9% of the mixture) is fully hydrated and will not expand in the steam curing process, and pigments may be added. The mixture is then pressed into moulds under great pressure and sent to cure by saturated steam over a period of 4–12 hours. The lime reacts with the surface of silica aggregate to form hydrated calcium silicates.

There is a drying shrinkage of 0.04% which is compensated for by expansion on wetting. The bricks produced are then a fine grained chemically bonded material, and their density gives them high strength which ranges from 20.5 to 48.5 N/mm^2.

Sand lime bricks have natural sand mixed with the lime whereas flint lime bricks have a large proportion of crushed flint, and are more textured in appearance. There is a natural range of colours from white to cream and pale pink, dependent on the sand and cements used, but added inorganic pigments (to BS 1014) can produce reds, browns, blues, purples, greens, blacks and yellows.

General durability

Sand lime bricks are durable and have a life expectancy of at least 40 years. They are reasonably frost resistant and show good general durability. They generally contain no soluble salts and are resistant to sulphates. They are more vulnerable to sodium chloride through salt spray causing surface erosion, and should not be used in marine situations.

Sulphur dioxide and carbon dioxide are gases present in the atmosphere that can affect calcium silicate bricks. Sulphur dioxide in contact with the bricks in moist air conditions can change the structure of the hydrated calcium silicate matrix converting it to a form of calcium sulphate, which can then quickly erode. This is more likely in polluted environments which have a high sulphur dioxide concentration. Carbonation can occur and change the outer skin of the brick to calcium carbonate causing case hardening with possible shrinkage. The order of shrinkage is very small and is rated in BS 5628 as being between 20 and 30% of the initial free moisture movement. There is natural drying shrinkage with these calcium silicate units. The only common fault with calcium silicate units is that they are liable to chipping along the arrises.

Common load-bearing and facing categories

The British Standard relating to this product is BS 187: 1978, revision January 1988. The compressive strengths vary from class 3 at 20.5 N/mm^2 to 4.5 N/mm^2 (see Table 4.20).

Table 4.20 Compressive strength classes, requirements and colours

Designation	Class	Mean compressive strength of 10 bricks not less than (N/mm^2)	Predicted lower limit of compressive strength not less than (N/mm^2)	Colour code
Loadbearing brick	7	48.5	40.5	Green
	6	41.5	34.5	Blue
	5	34.5	28.0	Yellow
Facing brick	4	27.5	21.5	Red
	3	20.5	15.5	Black
Facing or common brick	2	14.0	10.0	—

Source: Tables 2 and 3 BS 187:1978

As a specification guide, class 4 would be suitable for copings and class 3 for parapets, below damp-proof course level or freestanding boundary or other external walls. Class 2 would be suitable for bonded external walls, the outer leaf of a cavity, or internal walls.

The density varies from 1620 to 2400 kg/m^3 and an average density of 200 kg/m^3 will give a k value for thermal conductivity:

0.92 W/m K at 1% moisture content
1.24 W/m K at 5% moisture content

The differential relates performance to likely exposure and saturation, with 1% moisture content indicating an external sheltered position.

Like brickwork, the classes represent the original classifications which, in old units, stepped up every 100 lbf/in^2. The various grades are colour marked as indicated in table 3.

Movement joints

Movement joints are required at approximately 7.5 m intervals. The coefficient of linear expansion is approximately $8-14 \times 10^{-6}/K$ which works out at $1\frac{1}{2}$ times that of brickwork. Internal walls, which are not subject to dimensional changes by continued wetting and drying do not need movement joints incorporated as long as they are allowed to dry out fully before plastering.

Fire resistance

This property is comparable with clay and concrete bricks.

4.7 Concrete bricks and blocks

Concrete bricks are made from a wide range of aggregates which can also include the lightweight waste products of blast-furnace slag, clinker and lightweight pulverized fuel ash. The makers of some concrete blocks, such as forticrete, are finding that their traditional local sources of aggregate are diminishing, and the search for new sources means greater haulage costs. The aggregates are combined with lime and/or cement binders which may be moulded or formed into blocks by large-scale cutting machinery and then steam cured.

The natural colour of these bricks will be cement-coloured, although pigments will now give a range of browns and reds.

There is natural shrinkage in the making of these small-scale concrete components, in line with other cementitious material and concrete blocks should be allowed to dry fully and harden before the application of finishes. Tiles can be particulary problematic and de-bond if applied too soon after laying the blocks.

4.8 Masonry cleaning

See BRE Digest 280 for a summary on the cleaning of buildings generally. Water is the main cleaning agent but if this is applied too liberally it can hasten the corrosion of fixings and saturate some materials to an extent where efflorescence can become a problem, particularly in brickwork and stonework.

Chemicals are generally at best avoided as they often react with the substrate and cause even more problems and staining, giving the building a patchy appearance. Chemicals are also hazardous on site and should be minimized to avoid accidents.

Blast cleaning with an abrasive aggregate is common for heavy depositions but the grade of aggregate used should be carefully specified as it could involve irreparable damage to the face of buildings. As the aggregates involve fine air-borne particles of sand and grit, if containing free silica, they are now classified as hazardous to health under the Health and Safety Commission Code of Practice 1988. Free silica is also released during the cleaning process even if the abrasives are carefully chosen from the stone or material being treated.

Figure 4.7 Two houses by Colin St John Wilson, Cambridge 1968; (left) this project shows one of the first uses of forticrete blocks in this country with crushed Abergele limestone aggregate and waterproofed white cement; (right) interior with fair-faced finish. Pictures taken 1985.

Figure 4.8 The 'Times' shopping complex by Tadao Ando in Kyoto uses concrete blocks but they suffer from having iron pyrites as aggregates which give the small, dark, streak-like blemishes (iron sulphide).

Cleaning of brickwork

The Brick Development Association in their BDA Building Note 2, published September 1986, summarize some of the more common stains and defects found in brickwork that may need cleaning. The main strategies involve making sure that brickwork is not excessively saturated during construction, identifying the cause of the stain with expert help if necessary, and having trial treatments before carrying out cleaning to the whole of the affected area.

If chemical methods are used the brickwork should be thoroughly wetted so that chemical action is confined to the surface and not absorbed into the body of the brick.

White efflorescence can be removed by brushing away if the deposits are new and then sponging with clean cold water. Yellow or green efflorescence caused by the presence of vanadium salts will have to weather away naturally as inorganic acid solutions can cause them to darken. There are chemical methods of removal but they are not recommended.

Iron, in various shades of brown, can stain mortar and should be removed mechanically by filing or with an abrasive carborundum slip. A 5 or 10% hydrochloric acid solution is an alternative but it may be found necessary to try a variety of chemical treatments to find the most satisfactory one. Manganese staining is similar but darker in colour and responds to similar treatment.

Cement staining from mortar or concrete can be removed by 5% solutions of hydrochloric acid for light brickwork or 10% solutions for dark-coloured brickwork. Sometimes there is a washing out of lime from mortars or concrete which will respond to the same treatment if carbonated over a period of time, otherwise brushing may be successful if the deposits are fresh.

Deposits from pollutants, soot, dirt, etc.

These deposits may need removal with detergent solutions with the help of specialist contractors.

Organic growths, lichens etc.

These may be considered desirable in some circumstances and I have worked with one architect who advocated using sprayed diluted cow manure as an application to help obtain instant ageing of his new buildings or for repairs to existing rural buildings in the countryside. Growth is more pronounced in shaded areas which are more likely to retain moisture. (See BRE Digest 139.)

Figure 4.9 These traditional tiles at Matsumoto Castle, Japan, show the specialized types of tiles for ridge and gable details with fish sculptures as symbols of water protectors against fire.

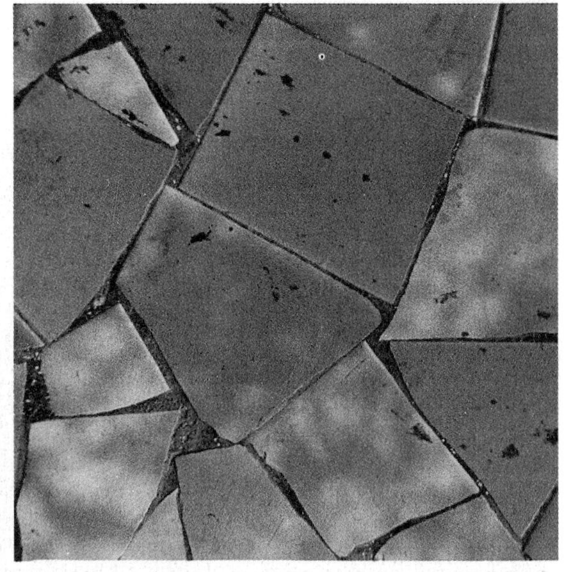

Figure 4.10 This Poissonerie in Paris shows the use of small stable tile fragments in providing a stable walling finish (already fractured into its small constituent parts) and used carefully to form precise lettering.

Figure 4.11 This building in Osaka, Japan, shows the use of tiles glazed with a high gloss metallic finish. Only when approaching closely is it apparent that the building is in fact tiled.

Vandalism

Vandals normally use various kinds of paint and it becomes a specialist operation to remove such deposits. Anti-vandal surface treatments make it difficult for spray paint applications to adhere to the surface. Reference should be made to BS 6270 *Cleaning and surface repair of buildings*.

4.9 Tiles

See *MBS: Finishes* for details on tiling.

References

1 Notes on Building Construction. In *Materials* (Rivington Series Vol. 3) Longman Green & Co. 1901.
2 *Designing for movement in brickwork* Brick Development Association Note 10 1988.

5 Glass

5.1 Introduction

Glass is a material that has been used for hundreds of years as a building material. It is one product which has a great capacity for development and, although the basic production of glass and its constituent materials has not changed, it now encompasses a wide range of products which have profound effects in the design of buildings. From its basic use as a window, defined by the medieval term *wind eye*, it can transform the behaviour and properties of walling systems. The section 'Glass as a medium for performance change' lists some major developments which are partly in production but which will become more easily available in the next five years. There are no British Standards yet which relate to these new materials. The Pilkington report, *Fenestration 2000*, published in 1989, summarizes future possibilities which also highlight new performance requirements needed, including the necessity for electromagnetic screening with regard to computer security applications. Currently glass is probably the most sophisticated material available to the building industry in terms of its wide performance range.

Designers have to think simultaneously about the light transmission and insulative properties of glazing and will probably have to think in terms of combining different glass types in order to give variable facade conditions. As glass is immensely durable it provides a weatherproof skin which is successful if it is well detailed within structural assemblies. Framing technologies are not currently keeping pace with developments in glass, particularly with regard to thermal performance, and specifiers are faced with greater reliance on performance specifications for product control.

Chemical composition of glass

Glass is a noncrystalline solid and does not have a regular repetitive three-dimensional structure of atoms, although there is local order. Traditionally the term refers to the inorganic glasses which are a mixture of complex silicates generated from the processing of soda lime and silica. If pure silica is used then the melting temperature is about 3,000 °F. When sodium is added the temperature can be reduced to about 1,460 °F. Calcium is often added as a stabilizer which helps to reduce chemical attack and dissolution by water. It also has the advantage of making the glass more malleable in working from a hot melt. In addition, to these main substances, properties can be altered by using any of the following:

Lead oxide May be added for cut crystal glass or radiation shielding.
Boric acid Controls the expansion rate of glass, making it heat resistant.
Flint and lead Can be added for optical lenses.
Aluminosilicate (Feldspar) glasses are high thermal impact materials used for laboratory and oven ware.

Recycled glass, known as *cullet*, may be added originating either from within the works or from the organized collection of glass. Metallic oxides may be added as modifiers. Manganese dioxide, selenium and cobalt oxide all help to de-colour the glass, removing unwanted impurities that affect colour. The oxides shown in Table 5.1 are added to control colour. Copper or gold as metallic deposits will give a ruby colour.

Glass does not crystallize on cooling, and sets before it is able to form the regular molecular arrangements characteristic of crystals. This is why it is described as staying in a liquid phase, and although the viscosity is very high, the liquid flow is minimal and the material is solid in appearance. It is apparent when replacing old glass that the glass has flowed, shown by the characteristic thickening of the glass at the bottom of a pane. It can sag in patent glazing and permanently deform into a curved profile, or

Table 5.1 Colour control

Material	Colour
Chromic oxide	Green
Cobalt oxide	Blue
Cuprous oxide	Red
Copper oxide	Blue–green
Ferrous oxide	Blue
Ferric oxide	Brown
Selenium	Pink
Uranium oxide	Yellow
Iron oxide and manganese oxide	Amber
Nickel oxide	Brown and purple

the ends of vertical panes of glass can show characteristic thickening. However, the term *glass* is applicable to the state of a material and is used to describe inorganic and organic materials, rigid materials which have also set after cooling without crystallizing and retain a random order characteristic of liquids.

Glass is then often described as a supercooled liquid, and it can also be classified as a ceramic due to its origin from the heating of inorganic materials, and the production of a material with ceramic-like characteristics. However, even this definition is inadequate as glass can also be produced from a gel which sets after the evaporation of a solvent, although needing further modification at relatively low temperatures (55 °C) to produce rigid silica sheets. Glass is also an elastic solid, and displays elastic recovery after the application and removal of stress.

Vitrification

Vitrification is the term used to describe the conversion of material into a glassy substance by fusion due to heat. De-vitrification can occur over a period of time, and the change from a liquid-like structure to a more ordered crystalline structure can eventually occur with shrinkage and ageing. Roman glass can show this characteristic change clearly, with a fine network of lines which are cracks initiating from crystallization. The glass is then extremely fragile. In glass-making processes the de-vitrification is also temperature dependent. If the glass is held and worked at a particular temperature, crystallization can occur, and the material then ceases to be a definitive glassy material. De-vitrification can also occur at ordinary temperatures, perhaps due to constant vibration or to gases such as carbon dioxide or sulphur oxide present in the atmosphere.

The material produced can be engineered so that it does set in a crystalline phase to give different properties. If alloyed with another material it can produce a glass ceramic which will be thermally stable. This is the same material used for cooker hobs and exploits materials known as *cervets* which have one component which shrinks on heating as it goes through a phase change. This counteracts the usual expansion of glass under heat and both elements will cancel to give zero expansion under heat.

According to the British Standard, BS 952 Part1:1978, *Glass for glazing*, transparent glass transmits light and permits clear vision. Translucent glass transmits diffused light without clear vision. Glass, a ceramic material which is already oxidized, is extremely durable and resistant to water and atmospheric gases. It can be affected by alkaline materials; rainwaters that run off from concrete, cement and plasters will damage the glass. In addition, there are some adhesives and chemicals used in cleaning masonry or paint which will also adversely affect glass.

History

Obsidian occurs naturally as an igneous rock, often of natural black glass. It was known as a material in the neolithic period and used for the making of arrow heads. There is evidence of its use at least 5,000 years ago. The Egyptians made glass in furnaces some 3,500 years ago and developed a system of immersing a solid core of sand into molten glass, letting it cool and harden and then scooping away the sand. Roman glass techniques were more advanced and, by dipping glass into a white glass surround, they could produce a design in sculptural relief by carefully chipping it away. Blown glass was not invented until approximately 100 BC. By increasing the centrifugal force in turning a blob of blown glass, it could be spun flat and this crude method produced glass which could be cut into small sheets generally known as *crown glass*. The central left over piece was known as a *bullion* and discarded, but often picked up for use in cottages. It is ironic that this waste piece is now mass-produced at some expense to fulfil a demand for a *cottage-like* appearance.

Mass production of glass became a possibility in the nineteenth century as mechanization of the process improved. Glass could be blown into large cylinders which had their ends cut off. They were then split into two pieces by diamond cutting and flattened gently under applied heat in a kiln. This was the first simple process in the making of *sheet glass*. The next development, which produced *plate glass*, involved pouring molten glass onto an iron table and then rolling it under a metal roller. The surface could be roughcast, rolled or polished, depending on the roller finish, and repetitive patterns are still produced in the same way. This process could also produce fluted glass if the table was profiled. Obscured glass was made either by grinding away the surface or melting powdered glass on the surface. It is now more likely for the glass to be acid etched or sand-blasted to achieve the same effect.

Since 1913 the process of glass-making has been on a much larger scale, with glass drawn up vertically into an annealing tower and then subjected to a rolling process for finishing and sometimes patterning. Annealed glass produces surfaces that are not absolutely flat and parallel and there may be some distortion. Alternatively, glass may be floated over a molten bed of tin which gives very accurate and parallel surfaces, a process used since 1959, invented and developed in England by Pilkington Brothers Limited. Ground and polished plate glass has largely been replaced by this method.

The cheapest glass is *horticultural glass*, green in colour as the impurities are not removed.

5.2 Design of glazing and choice of materials

BS 6262 *Glazing for buildings* replaces CP 152. The beginning of this standard gives a useful route for design by reminding designers of the following parameters:

- Initial sizing, location and likely cost of glazed areas
- Experience from other projects
- Actual requirements such as nature of lighting, thermal, sound, safety, security, fire, durability, wind loading and maintenance which may impinge on the original decision making.

Glass strength

Glass is a structural material and will have limitations on its span. Generally it is very efficient as a structure and, if supported on four edges, will have a span-to-depth ratio of 1:125 for single glazing, and 1:175 if insulating glass units are used. Reference should be made to figure 7 in BS 6262 which gives a thickness of glass relative to the size being used and the relative wind speed. Roughcast and patterned glasses are treated separately and are not as efficient, and wired glass is even poorer in performance. Double glazing units are the most efficient structures in wind, due to the rigidity of their box-like configuration. Toughened and laminated glasses will also give good performance.

Glass is subjected to continuing loads through wind pressure and has an ability to deform elastically and then return to its original shape. It fails in bending after the elastic limit is exceeded but, up to that point, is fatigue-resistant and can stand repeated loading, hence the durability of very old glass in windows.

Glass also has a very high theoretical strength but this relies on the glass face being free from notch imperfections. As an example of strength being related to surface quality, glass rods with slight imperfections were tested to destruction and achieved a breaking strength of 45 MN/mm^2. After sand-blasting, their strength was reduced to 13.8 MN/mm^2. After dipping in hydrofluoric acid to remove the flaws of sand-blasting they were re-tested and attained a strength which increased forty fold, to 1730 MN/mm^2. Even these strengths can be exceeded by glass fibres that have their surface coated and are quality controlled. They can produce tensile strengths up to 3,000 MN/mm^2.

Light transmission and solar radiation

As ordinary clear glass will transmit solar radiation, control is achieved by regulating the glass either to reflect radiation or partly to absorb it. If solar control is desirable various tints can be acceptable without great loss of light transmission. For glare control the light transmission will probably fall below 50% for effective control to be realized. This can be best achieved by treating the surface of the glass, known as *body tinting*, sandwiching a layer between glass by laminating, or enclosing an insulating layer. Surface treatments to the glass give a silver appearance and often a bronze transmission. Body tinting can come in tints of grey, bronze or green. These different glasses can be incorporated in double glazed or laminated systems if needed. It can be a far more effective way to design shading or blind systems to control glare, preferably external, to avoid any heat buildup on the internal side of the glass.

For exact calculation of solar gains in ordinary glass between 3 and 4 mm in thickness, a value of 0.87 is used for the total transmittance. As glass is used in different configurations, i.e. double glazed or incorporating some degree of control, the shading coefficient is the ratio of the total transmittance to 0.87.

Performance control by percentage of radiation transmitted is indicated for several configurations in Table 5.2. Table 5.3 gives typical solar radiant heat properties for a selection of glasses.

It is unusual to obtain glass that can give transmittance over 90%. If this is needed, the quality of materials used for the making of glass has to be controlled and must be made from pure white silica deposits of sand.

It is worthwhile designing systems which are responsive to annual variations. Glazing systems are often thought of as static installations, but they should be regarded as

Table 5.2 Performance control

Glass type (6 mm thickness)	Transmittance (%)
Clear	83
Body-tinted 75/61 green	61
Double glazing	71
Reflective double glazing	25

Table 5.3 Solar radiant heat properties

Glass type	Reflectance	Absorptance	Direct transmittance	Total transmittance	Shading coefficient
Single clear glass, 4 mm	0.08	0.08	0.84	0.86	0.99
Single or laminated clear glass, 6 mm	0.07	0.15	0.78	0.83	0.95
Body tinted, solar control	0.05–0.06	0.49–0.51	0.44–0.45	0.59–0.62	0.69–0.72
Laminated tinted inter-layer, solar control	0.04–0.07	0.21–0.81	0.15–0.72	0.36–0.77	0.41–0.88
Surface coated or modified, solar control	0.09–0.10	0.34–0.43	0.48–0.56	0.51–0.66	0.70–0.76
Laminated solar reflecting	0.12–0.48	0.35–0.55	0.09–0.50	0.22–0.62	0.26–0.71
Solar control double-glazing unit incorporating one 6 mm pane surface-coated glass	0.12–0.49	0.31–0.62	0.08–0.40	0.16–0.52	0.19–0.59
Clear double-glazing units having two panes, 6 mm	0.11–0.12	0.24–0.33	0.56–0.64	0.66–0.73	0.76–0.84
Solar control films applied to single 6 mm clear glass	0.14–0.50	0.30–0.66	0.14–0.50	0.22–0.60	0.22–0.70

Source: Table 6 BS 6262:1982

assemblages that give round-the-year control. Often they are designed for the worst conditions, but this can result in a loss of amenity and useful solar gain, particularly from clear skies in the winter months. There should be a conscious policy to allow occupants to vary systems as conditions change. There can be great variation in blind systems and external controls. The critical area for passive solar gain and control is within a six metre zone on the south side of buildings. Fixed blade systems designed for east and west elevations are more efficient if vertical rather than horizontal, and this will affect the architecture of buildings. There are also opportunities for design in the refurbishment of existing building stock which often represents 75% of the workload. Most ordinary glass will cut down the amount of ultraviolet content of light by about 50%, although there are special glasses manufactured which will allow 99% transmission of ultraviolet light. These glasses, having nil reflectance are quite difficult to see as planes of material, and may need special marking to prevent people walking into them.

Sound

As a material, glass does not absorb much sound. Its effectiveness as a barrier against noise is due to a combination of factors which include its overall thickness, whether it is laminated, and how it is housed within a glazing system. The size of glass is also important as a parameter, particularly due to its ability to resonate at different frequencies. Sound insulation can be improved by using double glazing, and there is an optimum for sound control in using an air space of about 200 mm.

Theoretical performance of sound reduction is not usually achieved in single glazing due to the resonance of glass. Laminated glass performs better in achieving sound reductions due to its ability to dampen sound through the intervening layers of materials such as layers of polyvinyl butyral (PVB). However, glass systems in general are better at insulating from higher frequencies. Most people have experienced the resonance of glass at low frequencies from traffic noise where vibration is very noticeable.

Table 5.4(a) and (b) gives sound reduction factors for a variety of glasses in decibels in relation to frequency and Tables 5.5 and 5.6 give some sound reduction index values.

Fire

Glass is generally regarded as non-combustible with regard to BS 476 Part 4 (apart from the polyvinyl butyral interlayer in laminated glasses) with a class 1 surface spread of flame. At high temperatures, although glass will not ignite, it will crack and so provide a pathway for fire. Wired glass has traditionally been used to hold glass together if it does fracture, and can be shown to have a certain fire resistance (which has size limitations). It cannot be looked at in isolation as framing conditions affect fire resistance. Performance can be improved by a laminated construction. For example, a sheet of Georgian wired polished glass (6 mm) with a polyvinyl butyral layer (0.38 mm) and clear float (3 mm) giving 9.4 mm overall thickness gives satisfactory performance for stability and integrity, under BS 476 Part 8, for 60 minutes.

Some laminated glass constructions (such as *Pyrobel*, manufactured by Glaverbel, or *Pyrostop* by Pilkingtons) incorporate intumescent interlayers which expand on heating. These cause the glass to become totally opaque but still give complete protection with ratings that increase fire resistance from 30 to 60 minutes, depending on the thickness of the laminated construction. Use is normally limited to interior partitions as the laminate can be degraded by external weathering or solar radiation.

Borosilicate glasses are used for their general fire resistance which can be up to 2 hours dependent on the glazing details used. Their ability to resist thermal shock

Table 5.4(a) Sound insulation of glass — single glazing performance

Centre frequency of third-octave band (Hz)	Sound insulation for a given glass thickness (dB)						
	4 mm	6 mm	6 mm (laminated)	8 mm	10 mm	12 mm	12/14 mm nominal (laminated)
100	20	22	22	23	24	25	25
125	20	22	23	24	25	26	27
160	21	23	22	25	26	27	27
200	22	24	23	26	27	28	29
250	23	25	26	27	28	29	32
315	24	26	28	28	29	31	32
400	25	27	30	29	30	32	33
500	26	29	31	30	31	33	34
630	27	30	32	31	32	33	34
800	28	31	34	32	32	32	33
1000	29	31	35	32	31	29	33
1250	30	31	34	29	27	26	32
1600	31	28	32	26	27	30	36
2000	29	25	30	28	31	34	40
2500	26	27	32	32	35	37	43
3150	23	31	37	35	37	40	46
Mean (100–3150 Hz)	25	27	29	29	30	31	34

Source: Table 10 BS 6262:1982

Table 5.4(b) Sound insulation of glass — double glazing performance

Centre frequency of third-octave band (Hz)	Sound insulation for a given thickness of glass/air space/glass (dB)							
	6/12/6 mm	10/12/6 mm	7 nominal (laminated) /12/4 mm	6/100/4 mm	10/100/6 mm	10/200/6 mm	6/200/6 mm	6/200/6 mm
100	21	22	23	23	26	33	32	32
125	27	28	25	27	29	37	35	35
160	27	28	23	30	32	39	38	37
200	23	24	22	33	34	41	40	38
250	25	25	23	35	36	42	42	39
315	28	29	27	37	38	44	43	41
400	29	30	29	39	40	45	45	43
500	31	31	32	41	42	47	46	44
630	32	32	35	43	44	48	47	45
800	33	33	38	45	45	49	48	46
1000	34	34	40	46	47	50	49	46
1250	33	34	41	48	47	50	50	46
1600	31	33	40	49	48	50	49	44
2000	27	34	39	49	50	51	44	39
2500	29	35	39	50	52	52	47	41
3150	34	36	43	50	54	54	52	46
Mean (100–3150 Hz)	29	31	32	40	42	46	44	41
Sound absorbent in reveals	No*	No*	No*	Yes	Yes	Yes	Yes	No

* Doubles-glazing units.

Source: Table 10 BS 6262:1982

Table 5.5 Sound reduction index values for single glazing

Thickness (mm)	Mean sound reduction index (dB)
3	18
4	20
5	22
6	23
8	25
10	26
13	28

Source: Table 18 BS 6262:1982

is due to the sodium borosilicate mix having zero expansion under the effect of heat. These glasses are thick, commonly 6 mm in thickness, and usually have quite limited sizes as the maximum sheet size made is 2.00 × 1.2 m. Low expansion borosilicate glasses have great fire resistance but they have to be relatively thick, and would need to be at least 20 mm thick to achieve 60 minutes fire resistance to BS 476. They have a greater fracture toughness than ordinary glass. The glass will contain at least 5% boric oxide and will reach a softening point at 600 °C. *Pyrex* is also a trade name for this glass composition, which is also used in the making of fibres.

There are new glasses which use phase transition behaviour to produce glasses that will not expand and crack under heat. These panes should be clearly marked or identified in order for them to be replaced with the same kind of glass specification in the event of breakage, as they have to be ordered to size.

Glazed elements have to meet the requirements of structures of which they are part, if those structures have to attain a certain fire resistance. Their use has to be looked at carefully if they form part of fire routes. See BS 5588 for specific advice in case the percentage of glazing may have to be limited. Glazing cannot be looked at in isolation from its supporting structure and pane size may be limited (see BS 476 Parts 8 and 22). If this guidance is inadequate, advice should be sought from the manufacturer. It cannot be assumed that if a particular sized pane in a specific supporting structure reaches a certain fire resistance, a multiple assembly of the same components will perform in the same way.

Beads are important in sizing and the edge restraint of glass in conjunction with providing temperature control at the edges of the pane. This is the factor which determines whether the glass pulls out from the frame under fire conditions. Insulated glasses do not fail in this way and the actual body of the glass is affected directly by heat.

It is possible to meet a fire resistance period of over one hour with glazing if sized, specified and detailed correctly.

Glass cleaning and effects of water

Contrary to popular belief, most glass is not self-cleaning and should not be put in situations which are inaccessible. Glass can be degraded by contact with water, acid and alkali solutions, as well as salt solutions and gases. There can be distinctive loss of clarity with glass over time which is a visual indicator of corrosion. McColm[1] identifies three major effects which include:

- Reaction with the corrosive agent to form a new compound surface layer.
- Preferential leaching of components of the glass having a leached surface layer.
- Total dissolution, continuously exposing fresh glass.

There have been periods when it has been fashionable to use patent glazing or detail vertical glazing, with a high level raked detail over the top. These are particularly difficult to clean and, apart from directly reducing light transmission, as they are not cleaned, the surface of the glass becomes roughened as a new porous surface film is built up which is then even more likely to catch deposits

Table 5.6 Sound reduction index values for double glazing

Thickness (mm)	Mean estimated sound reduction index for a given air space (dB)										
	0 mm	10 mm	20 mm	30 mm	40 mm	50 mm	75 mm	100 mm	150 mm	200 mm	250 mm
3	23	24	25	26	27	27	28	30	31	32	33
4	25	26	28	28	29	30	31	32	34	35	36
5	26	28	29	30	31	31	33	35	36	37	38
6	27	29	31	32	32	33	35	36	38	39	40
8	29	31	33	34	35	35	37	39	41	42	44
10	31	33	34	36	37	37	39	41	43	45	46
13	32	35	37	38	39	40	42	44	46	48	49

Source: Table 19 BS 6262:1982

and deteriorate further. Water which remains on the surface of glass causes sodium ions to leach out in solution, changing the pH of the water solution to a more alkaline state assisting further degradation. Corrosion inhibitors can be added to glass to help prevent this phenomenon.

Acid solutions can neutralize alkaline reactions and help form thin but protective surface layers. Detergents can be classified as salt solutions which can form complex compounds with metal ions in the glass. Depending on their composition, if they have a high pH (11.9) they can cause alkali attack on SiO_2, but a lower pH can induce silica deposition. Glass cleaning should therefore relate to the type of glass and also its situation. Mild detergents are recommended and if sponges and neoprene-bladed tools are not available, using newspaper to absorb and polish off soaped surfaces is a very effective way of cleaning glass.

There is a risk of some degree of solubility with glass and this is increased if modifying oxides are added. A porous surface film is produced in static testing with water lying on the surface which encourages further degradation. Alkaline solutions are the most aggressive and acidic solutions will neutralize further reactions by helping the formation of a strong protective layer. The alkaline degradation phenomenon can be seen if ordinary tap water in hard water areas is left in drinking tumblers for any length of time. The surface layer formed is tenacious and difficult to remove. Strong detergent solutions can cause silica deposition. (See McColm[1] page 197.)

Ordinary glasses are susceptible to chemical attack by virtue of the mobilization of ions within the material. This mechanism can be used to advantage in manufacturing processes where ion exchange can allow for the integrated deposition of silver ions, which are similar in size to sodium ions. Wet glass can also be stained if in contact with iron or manganese.

Once the reasons for glass degradation are known, it becomes possible to specify protection, and specifiers should check on which system is employed depending on glass usage. Thin ceramic films can be deposited to provide an inert layer. A silicone surface treatment can also be applied; in a controlled reaction this produces a surface layer of orientated methyl groups which can repel water.

5.3 Types of glass

Glass is categorized by its method of manufacture.

Annealed glasses

Annealed flat glasses are subdivided into float or polished plate glass, sheet glass, cast and wired glass. All annealed glasses are cooled slowly and are stress relieving.

Float and polished plate glasses are of the best quality, the surfaces are parallel to each other and there is no distortion. Float glass is literally floated over the surface of molten metal and is made in thicknesses up to 25 mm, now superseding polished plate glass. Plate glass is made from roughcast glass which is then mechanically finished to a high standard and can be obtained in thicknesses of up to 38 mm. Both methods of manufacture can allow for tinting in green, grey or bronze, but in limited thicknesses up to 12 mm. Float glass can incorporate metal ions to provide solar control or alternatively achieve protection by the provision of selective coatings.

Sheet glass is manufactured in a continuous process with fire-finished surfaces. The surfaces are not parallel as in the cast or floated processes with their consequent distortion of image. Sheet glass can also be obtained body tinted in green, grey and bronze. Cast glasses are manufactured using a rolling process and the rollers can incorporate patterns. The deeper the pattern, the greater the internal reflection, and the glass is consequently more obscured. They can also be obtained tinted. In this process it is easy to incorporate a mesh to make wired glass.

Toughened glasses

Toughened and laminated glasses are usually specified in response to safety or security requirements. See BS 6206 for classifications generally with regard to types of glass and their properties with regard to safety.

Annealed glass is taken through a further stage in processing which reheats the glass and then cools it quickly using jets of cold air. The outer surface cools first and gives a surface which is highly compressed, and a body of glass in the centre which is in tension. The glass is consequently prestressed and is far more difficult to break initially. It attains a working strength which is between three and five times stronger than ordinary annealed glass. This glass is usually referred to being *fully tempered*. Once broken, it will shatter into small pieces which are less harmful. If holes need to be made in the glass this should be carried out when the glass is in the annealed state and prior to it being toughened. As it cannot be cut or have any part of it remade after processing, the exact size has to be ordered. All types of annealed glass can be toughened. In addition, coloured ceramic enamels can be fired onto the surface of glass while it is being toughened to provide a cladding material. There is an intermediate process which uses heating and cooling but the cooling is not so extreme and the glass will just double the strength of ordinary glass. This process may be referred to as *heat strengthening*.

There can be spontaneous failure through the swelling of impurities present in the glass. Nickel sulphide crystals might increase in volume by as much as 3.9% over time.

The microcracks formed can shatter the whole of the glass area.

Chemical toughening Chemical toughening can be achieved by exchanging ions at the glass surface. Larger potassium ions from a molten salt bath migrate into glass, replacing the smaller sodium ions. This produces surface stress with the same effect as pretensioning. As the interchange happens on the surface, affecting only a 20 micron depth, the toughening mechanism can be weakened locally by scratches and defects.This process is often confined to small-scale components and optical glasses. One main advantage is that glass toughened in this way can still be cut to size after the process.

Laminated glass These are of sandwich construction usually with a bonded layer of polyvinyl butyral between two panes of glass (known as three ply) that prevents forced penetration and helps to keep the glass from disintegrating on impact. If broken, pieces of glass will still adhere to the interlayer, which may only be 0.38 mm thick. For high security, additional layers of rubber (which can also increase in thickness) and glass can be bonded and will resist impact by pickaxe and, if over 19 mm in thickness, will resist penetration by bullets. As the glass is laminated, different panes can be chosen to give additional properties, such as solar control using reflective or heat absorbing mechanisms. Manufacturers should be consulted when glass is being specified as the range of performance can extend from general security to withstanding firearm attack as well as explosions.

Wired glass Wired glass will hold glass together after it has been broken by fracture from impact damage or by fire. If 6 mm in thickness and glazed in openings with the right choice of materials (which do not exceed 1.6 m^2) it can achieve 90 minutes fire resistance.

Special glass products

Hollow glass blocks All properties of glass blocks are dependent on the individual product chosen from particular manufacturers and it is difficult to generalize. BS 1207 and CP 122 can be referred to for guidance. The blocks are made by pressing glass into a mould. Two moulded units are then placed together and fused to form a hollow form. There are two main sizes:

240 × 240 × 80 mm
190 × 190 × 80 mm

It is possible to obtain 300 × 300 mm units and also 100 × 100 mm units. Rectangular shapes and complete circles are also available.

Figure 5.1 Prototype modernization at Bishopsfield Housing Estate, Harlow (1994) by Prof. Florian Beigel of the Architecture Research Unit at the University of North London. The façade design uses fixed clear glass and transparent insulation; areas for ventilation are solid openable panels.

The process allows for a number of patterns to be printed and they can be in coloured glass. They are used chiefly for their obscured vision quality, giving light but privacy. They can achieve a light transmission of up to 85% if the smaller size clear block is used, although 75% is more usual. Lenses are also made, mostly for roof lighting and pavement lighting and are pressed in form. Glass blocks can be used for whole walls but will often need to be constructed against temporary timber formwork to allow for the incorporation of steel reinforcement. The panel size is limited with maximum dimensions not exceeding 6 m (overall area 18.5 m^2). Dry fix systems may limit panelling to one hundred blocks (U M Hills Glass recommendation). Glass blocks can only support their own weight and cannot be considered as load bearing walls.

Airborne sound reductions can be obtained using glass blocks of between 40 and 49 dB. Thermal transmittance varies but can be between 2.5 W/m^2 K and 3.20 W/m^2 K. Fire resistance for half-hour and one-hour periods can be obtained depending on the glass blocks used and if the method of construction is *in situ*. Blocks are generally regarded as non-combustible with a Class O fire rating. Solar gain is generally negligible if the angle of the sun's rays is below 30°, and 60% of the sun's heat energy is transmitted when the sun is normal to the panel's surface (i.e. on the horizon). (*Source*: U M Hills Glass) Due to the three-dimensional nature of the block, the width of the mortar joint acts as a shading device, giving cut off to the sun's rays and can be imagined as a deep grid. *In situ* constructions benefit from having synthetic rubber in liquid added to the main mortar mix to increase adhesion. Expansion should be allowed for by detailing flexible joints at the perimeter of the panel.

PC Glass Blocks (Pittsburgh Corning UK)
U value 0.51
R value 1.96

These products can also incorporate fibrous glass infills which affect the light transmission properties and also improve the thermal resistance by approximately 5%.

Solid lenses Solid glass blocks are produced that can be used as pavers (often 75 mm thick). Their size is small to allow for closer mortar joints to prevent slippage. Some lenses are also produced that are prismatic and can give directional light.

Glass channels These sections used to be popular for making wall assemblies and can be used for single glazing or lapped or butted to create double glazed walls. They can also be used for roof lighting. They are imported with channel sizes of 262 mm in width and with a depth of 41 mm. They can be produced with embedded wires and have been used more recently as long shelving units showing how their structural strength can be exploited for a use that was quite unintentional. Jointing systems which used polysulphide joints can now be far neater with silicone sealants. They are made from cast glass and give a translucent appearance.

Coloured glass As an alternative to using body-coloured glass, a range of products uses layers of colour coat bonded to the underside protected by a clear body of glass. Colour ranges can be vast and matched to the ICI colour dimension system as used by Chelsea Artizans known as the *Diamond Chromatics* system. Colour coatings can be thick, between 80 and 100 microns. Computerized colour control compensates for any reflected green tint of the glass. Colour application can be mixed allowing for graphics and logos.

Ashlar glass linings

There is a roughcast glass, 10 mm thick, which is produced for industrial use particularly as linings in industrial processes, providing a durable surface which does not need any maintenance. It is normally set in rubber latex cements, often on sloping wall surfaces that are masonry or steel framed.

Opacity through mechanical treatments

Sand-blasting Compressed air is used to blow sand particles against the glass surface. The consequent roughening will scatter light and give a translucent effect. Stencils can be used to control designs. As the surface is damaged this weakens the glass and it can suffer a reduction in strength of 50%. This should be taken into account when dealing with large pane sizes which will not have the expected span:depth structural ratio of that glass thickness.

Acid etching Etching can produce a wide range of surface from heavily frosted to a polished but translucent appearance. The main chemical used is hydrofluoric acid and its area of use is controlled by the *resist* materials of the first coatings.

Enamelling

Colours can be fired onto glass surfaces. There are additional benefits as the surface of the glass can then be thermally toughened through this additional process giving a durable panel.

5.4 Structural glazed units and wind loading

Most glass is used in panes or sheets to admit light and solar radiation. Every window unit has to be designed to ensure that the glass size is adequate in terms of strength. This is in respect of glass thickness and type with regard

Figure 5.2 Etched glass wall to restaurant, sea front, Barcelona. The etching gives privacy, the name of the restaurant and an indication that it specializes in sea food. The whole glass wall can also move on rollers.

to the possible pressures that may be exerted on the glass, not only from external wind pressures but also internal pressures in air conditioned buildings. Studies of wind pressures on buildings by the BRE have not been revised since 1976 (CP 44: *The response of glazing to wind pressure*). This validated the pressure coefficients used in the United Kingdom glazing code at that time for peak effective pressures. (See British Standards Code of Practice CP3, chapter V, Part 2, 1972: *Wind loads*, revised 1989.) This has now been replaced by BS 6399 *Loading for buildings*, but the Part 2 Code of Practice for wind loads is still being revised.

Wind speed is converted into dynamic pressure by using the formula:

$$q = kV_s^2$$

The design wind speed V_s is then calculated from

$$V_s = S_1 \times S_2 \times S_3$$

V is the basic wind speed taken from the map of the United Kingdom (refer to Figure 1 BS 6262 Basic wind speed) and then modified by wind speed factors which include topography, ground roughness, building size and height and probability of exposure to exceptional storms. (Refer to Figure 3 BS 6262 Conversion chart for wind speed and dynamic pressure.)

BRE Digest 346 (1989) deals with *The assessment of wind loads*. Gust speeds can be obtained from the BRE computer program *Strongblow*.

Figure 5.3 Stansted Airport, Foster Associates (1991). The opaque glass eliminates the usual problem of glare at high levels of fully glazed walling systems.

There is a simplified method described in BS 6262 for establishing design wind pressures which uses the basic wind speed obtained from the map of Britain and combines this with the information given in Table 5.7.

Edge clearances

All glazing should allow for edge clearances which will vary according to the size and thickness of glass, avoiding direct contact between the edge of glass and the frame. This will vary from 3 mm for most types of glass, which is up to 12 mm in thickness and up to 2 m in its maximum dimension, to 100 mm for a laminated glass which has an overall thickness of 30 mm which is over 2 m in its maximum dimension. This is in addition to the requirements of rebates for wind loading and will result in a geater rebate depth.

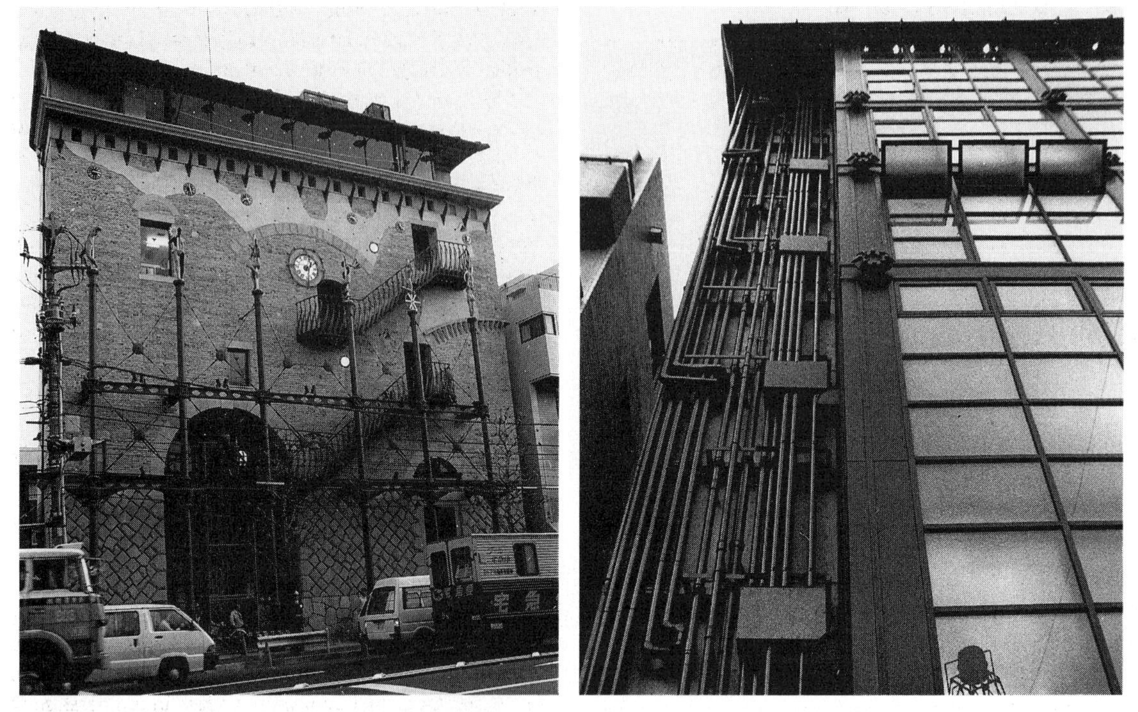

Figure 5.4 'The Wall' by Coates Branson (1990): (left) façade; (right) rear view. This slim office block has the entire rear wall in opaque glass. The view would have been on to an untidy scramble of back buildings and this device gives an insular quality to the interior with a high quality of light.

Table 5.7 Ground roughness categories

Ground description	Category
Long fetches of open, level or nearly level country and all coastal situations	1
Open country with scattered windbreaks	2
Country with many windbreaks; small towns; outskirts of large cities	3
Surfaces with large and frequent obstructions, e.g. city centres	4

In addition, rebate depths should increase as wind loading increases. Table 12B of BS 6262 will give a guide where the range is from a 9 mm depth at a wind loading of 1000 N/m^2, to an 18 mm depth for a wind loading of 3000 N/m^2.

Source: Table 1 BS 6262:1982

5.5 Insulating glass units

These units are often hermetically sealed and made under controlled conditions. Air or alternative gas-filled spaces (often argon) are dehydrated. They can be made from straightforward annealed glass or toughened and can have a range of finishes for solar control. Pilkingtons have a system using *Kappafloat*, a low emissivity coating, as the inner pane which reflects out any heat absorbed by the outer pane. This combination can reduce the U value of ordinary double glazing from 3 to 1.3 $W/m^2 K$. This reduces heat loss from long wave radiation but still allows the transmission of solar heat. The glass enclosure can be made in a variety of ways from an integral glass structure which is welded together or uses a metal section which can also be welded. If metals are used in construction they should be corrosion resistant. Alternatively there are systems that involve spacers fixed with adhesive or mastic and incorporate a desiccant which will absorb moisture without liquefying. The European Union of Agreement has a directive for standards of these components which restates generally the kinds of design parameters that should be considered, for example, wind force and atmospheric pressure, temperature, fire resistance and requirements for water, air and snow-proof seals. Table 5.8 gives the thermal resistance related to the likely width of cavities.

There is very little to be gained from specifying over 12 mm as a cavity width. Cyclical testing is used with temperature extremes between −15 and +55 °C with 95% humidity at the highest temperatures. Particular attention is paid to organic bonding sealants and their ageing in respect of solar radiation.

The most common sealants include polysulphide, polyurethane, silicone, epoxy-polysulphide or butyl. The

Table 5.8 Thermal resistance values

Width of cavity (mm)	Thermal resistance in vertical wall (m^2 K/W)
6 and 7	0.11–0.13
8 and 9	0.12–0.14
10 and 11	0.13–0.15
12 and 13	0.14–0.16
14 and over	0.15–0.17

silicone family of sealants are useful in that they allow the outward passage of water vapour but will resist the inward passage of larger liquid water molecules.

The life expectancy of sealed glazing units is not as high as single glazing. Gases and moisture vapour may eventually diffuse through sealing compounds. It is vital that full drainage systems are incorporated below the base of these glazed units, unless special silicone sealants have sufficient surface area to liberate any trapped water vapour through natural evaporation. For drained systems the minimum rebate height should be 22 mm, and for ventilated and bedded systems the minimum rebate height should be 18 mm. For complete advice on the detailing of these sealed systems, refer to the Glass and Glazing Federation (GGF) publication *Glazing techniques for insulating glass units*, March 1988.

The whole unit should be detailed in the rebate to allow for positioning with load bearing wedges or setting blocks which will evenly distribute loading from self-weight or external pressure. The best materials for these blocks are described in the *Manual of the Glass and Glazing Federation*, section 4.2 as of 'resilient, nonabsorbent compatible materials' and likely to be sealed hardwoods, extruded uPVC, plasticized PVC (BS softness number 35–40) or neoprene (shore A hardness number 80–90). In addition, there are location blocks to prevent movement between the unit and its surround, and distance pieces which are positioned vertically between glass and the side of a rebate or bead which prevent movement in the glazing compound under applied wind loading. Packing will also allow for the differential movement between glass and frame. There will be different coefficients of expansion between glass and its framing substructure whether in glass or metal.

There can be some visual distortion from multiple surface reflections. As glass attains higher standards of manufacture and is perfectly flat, a distortion phenomenon can arise known as *Brewster's fringes* which is an example of the interference of light similar to those seen in the toughened glass of windscreeens. This occurs when direct sunlight meets directly reflected light from a surface of the insulated glass, the wavelengths of light cancel each other out (they would be stated as being 180 degrees out of phase). It is ironic that this phenomenon is indicative of the very high quality of the parallel nature of the glass surfaces.

In choosing double glazed units, care should be taken to match the exposure rating of the site with the rating of the unit. For the best performance, glazed units should be fixed in rebated frames with beads. Stepped insulating glass units have a low exposure rating as sealants are not protected and drainage details are nonexistent, and rely on a putty fronting requiring maintenance. As these units are comprised of a number of parts, if there are problems the GGF have produced a standard procedure for dealing with claims against warranties of sealed unit manufacturers.

Insulating glazing is now much extended as a range with the development of insulant materials that can be incorporated into a sandwich-like structure. They use gels, honeycombs and fibres usually made from a range of polymeric materials.

Rebates

There are considerations in specifying insulating glazing with respect to the rebating of the glazing system. Here manufacturers' recommendations should be followed. It is likely that there will be a great variant in temperature if inadequate rebates are used, and thermal shock to the glass could cause breakages, especially if temperature differentials are of the order of 25–30 °C. There are particular problems in installations at altitudes of over 1000 metres, where high solar radiation is present with cold air temperatures. Ideally, frames should be designed with thermal breaks.

5.6 Frameless assemblies

Frameless assemblies

There is now a range of examples from the first part of the twentieth century with early developments by Jean Prouve where whole walls are made of glass, for example in the Citroen showroom in Paris. Frameless assemblies demand high structural performance from materials as stresses have to be borne directly by fixing components, glass and sealants. In these situations the full design and specification has to be a joint effort between architect, engineer and specialist contractor.

The structural systems of support become of interest when the normal structural strength of glass is exceeded, and the usual orthogonal framing systems also reach their limits in respect of the likely structural strength of mullion and transom sections. The greatest consideration is a deflection applied to wind loadings. Butt jointed details using silicone are more common and the sizing of the glass and its thickness has to be established on assuming possibly only two edge supports. Glass fins can be used as structural stiffeners and they will normally be at least 12 mm in thickness. As the whole window wall system will be subjected to negative wind pressure, the adhesive joint will have to perform under loading conditions of suction and not compression. The British Standard BS 6262 gives guidelines for assemblies up to 5 m in height. Assemblies which exceed this dimension will have to be sized as separate engineering designs, usually between the manufacturer and the structural engineer. Refer to tables 39–41 in BS 6262 which help determine the glass thickness, and the fin height and width.

The tops, bottoms and sides of assemblies are usually housed into the substructure and are concealed. Any further division will be silicon sealed and stiffened with glass fins or other stiffening structures. Patch fittings might be used as connectors. These are deceptively simple systems which often involve the hanging of one pane of glass from another. Engineers such as Peter Rice have extended the structural vocabulary of glass systems by their use not only of vertical stiffening members but also by the design of horizontal tension trusses to resist wind loading and internal collapse.

Due to the development of adhesion technologies using silicone sealant systems, it is possible to make glass walls which are frameless. The silicone joints are now strong enough to transfer the dead loads of the glazing material and live loadings from wind loads to other assemblies which are remote and not fixed directly to the glass structure.

Alternatively, patch fittings may be used to hold glass at individual points instead of using continuous frame assemblies. Using the *Pilkington wall* structural glazing system as an example, it is likely that 12 mm armour plate glass, horizontally tempered, is used in these situations. Depending on the height of the structure, stiffening glass mullions of 19 mm armour plate glass may be needed. Glass panels are pre-drilled in their corners prior to toughening and stainless steel bolts connected to plates or patches which can be stainless steel or high tensile brass (to BS 2874); sized at 165 mm × 165 mm total width across two sheets. These patch fittings can have a variety of metal finishes, i.e. satin or polished, or black in the case of steel, or chromed in the case of brass. All the small-scale components have to be durable in terms of thermal and structural stability. Separating fibre gaskets to BS 2768 are used in conjunction with *Nylatron* polyamide bushes. Flexible silicone sealant is used between glass panels (to BS 5889). Suspended systems can be designed up to 23 m and ground-based systems to a height of 9 m. There is unrestricted length, although this will be dependent of wind conditions and limitations of the building structure.

Horizontally tempered glass panels will be a maximum of 4200 × 2000 mm at 12 mm thickness (weight 30 kg/m^2) and mullions will be a maximum of 4200 × 1000 at a thickness of 19 mm (weight 48 kg/m^2).

See *MBS*: *External components* Chapter 6 dealing with glazing without frames for construction details. The planar glazing system from Pilkingtons uses smaller fittings which are fixed back to vertical frame members.

5.7 Mirrors

Mirrors should be thought of as not simply reflecting images, but as devices that can also change the nature of the lighting in a space, and the direction of the lighting in a space. If this does become a prime consideration mirrors should be placed as near as possible to the primary light source for maximum efficiency. The greater the expanse of mirror the greater the illusion of space.

Mirrors work by having a deposition of metal on their rear surface. Most domestic mirrors work by having a deposition which allows for aluminium to be evaporated onto the glass surface. Spray applications of silver nitrate and tin chloride are also used. Subsequently the back has to be coated to protect the thin metallic layer, which may be either a resin or paint. For more permanent results an electroplating of copper could be specified which may have a further enamelled and subsequently stoved finish. As the coating is applied to the back of the glass, the thickness of the glass will give approximately a 4% internal reflection which can show as a double image in the glass if viewed from an angle. This is not usually a problem in most situations. Optically accurate mirrors used for astronomy will have the deposition on the face of the glass.

One problem arising out of the process of making mirrors is their fixing. Adhesion bonding will rely on bonding to the coatings to make the mirror and not the glass itself, and this also applies to the wall finishes. An air gap of 3–5 mm is recommended between mirror and wall to protect the mirror coating against the effects of condensation (especially in bathrooms). Adhesion bonding normally uses pad fixings which are sized and spaced according to the weight and size of the glass used. Mirrors can also be fixed mechanically but walls should be true, or packing washers should be used to prevent curvature and deflection which could cause unacceptable stress. Any holes drilled in mirrors should be finished to prevent stress fractures developing. Washers of rubber or nylon should be used both sides of the mirror and not overtightened. *Silvering* is used rather loosely as a term to describe mirroring which can be achieved with different metals. The GGF publication *Recommendations for fixing mirrors* has some particularly useful details for junctions.

If mirrors are to be fixed on to uneven backgrounds, there should be a minimum bedding thickness of 5 mm. As the background of the mirror may be silver, it could be affected by sulphur (due to the potential of the reaction between silver and sulphur to form sulphides — metal ores are commonly sulphides). Sulphur can be found in clinker blocks and levelling cements. There should be direct barriers against these unfinished materials; polythene is thought to be adequate. Alkalis in cements, plasters or concrete can also attack the backing paint of a mirror, and water vapour can also degrade the backing. In these cases a gap of 3 mm will be adequate for separation protection.

One-way mirrors

This coated glass gives a reflectance of 38% with a silver appearance and is normally only suitable for internal applications. The lighting levels have to be different on either side of the glass and the ratio used is normally 7:1 of the observed area to the observation area. Cleaning has to be carried out carefully as the coating could be damaged.

Diffuse reflection glass

This works by a texturing of the glass which has a nominal thickness which is very thin at only 2 mm. There are also limitations on the sizing of 1830 × 610 mm or 1850 × 1240 mm.

5.8 Glass working

It is worth referring to BS 952 Part 2: 1980 which details the kind of cutting work that can be undertaken on glass. There is a range of cuts that can be made in the surface, and complex three-dimensional shapes, such as finger slots, can be incorporated. Round or oval depressions called *punts* in the trade can be specified and will change the quality of vision. Glass edges always need attention and they can be rounded, bullnosed, squarecut or bevelled, and all of these shapes can be ground, smoothed or polished. Double bevels can be used to reduce the apparent thickness of glass doors for example, and form three-dimensional frames around doors and windows. Traditional scalloping and fluting, more often seen in mirrors can be undertaken.

Methods for obscuring glass can be more inventive than purely ordering white or obscured glass. There are different methods for making obscured glass which include sand-blasting, grinding or acid etching. Sand-blasting can have different grades; fine, medium or coarse. These all give white effects to the glass. The glass can be ground to give a more subtle finish which is not so white as sand-blasting. Acid treatments can also give a range of finishes and different degrees of obscuration.

Because glass is such a fundamental material in building, it literally acts as the *wind eye*, the Middle English descriptive term which was more accurate then as it primarily admitted air, being covered when ventilation was not needed. The name is now corrupted to *window*, an element which protects as well as gives vision and fresh air. Due to its nature it is an important part of the building envelope and is in itself a modifier of the internal to external environment. As energy management of buildings becomes more important, the sizing of windows, their construction and their ability to modify the quality of the interior environment in respect of passive solar gain and heat loss becomes more critical. As a window element glazing is subject to an international standard ISO 9050: *Glass in building. Determination of light transmittance, solar direct transmittance, total solar energy transmittance and ultraviolet transmittance, and related glazing factors*. This standard sets out the methods of calculation of all the above parameters.

Although the domestic market uses chiefly transparent glass, in commercial and public buildings specifiers are becoming used to choosing from a wider range of products. Often these products are used to modify an external glass skin when the extent of the glazed skin itself could be modified. To ameliorate the effects of passive solar gain, glasses are often specified which have the effect of blocking out passive gain at the expense of the occupants. For example, if a metallic composite glass is used for solar control it will also have the effect of giving a permanent grey sky to the occupants. These glasses also block out selective wavelengths of the light spectrum. This is not beneficial to people who may suffer from decreased resistance to colds and other ailments as a result of losing wavelengths in the ultraviolet light regions.

Depending on how glass is processed it can possess a range of properties and differing transparency. Often in different situations advantage is not taken to specify different glass types. For privacy or to obscure external features, buildings that use sand-blasted glass which will let in a diffuse quality of light will be more appropriate as a solution. Windows can be designed to give borders of clear glass or clear centralized vision with perimeter translucency.

Glass bending

Flat glass is made first and then re-heated and cooled around metal moulds. Glass domes are categorized as either rectangular (barrel-shaped) or circular. (Refer to BS 952 Part 2:1980 for limitations on the shapes that can be produced.) Table 5.9 lists some typical dimensions

Overall limitations are 5000 mm on total height, 914 mm for the depth of curve and 3500 mm for the girth of the curve.

Special shapes can be produced to order and can be made with Georgian wire to satisfy BS 476 Part 8: 1982 for roofs, and also BS 476 Part 3: 1958 for external fire exposure.

Figure 5.5 Finsbury Health Centre, Lubetkin (1936).

Table 5.9 Typical dimensions of rectangular and circular glass panels

Rectangular (mm × mm)	Circular (diameter in mm)
600 × 600	600
750 × 750	750
900 × 750	900
900 × 900	1050
1050 × 1050	1200
1200 × 600	1350
1200 × 900	1500
1200 × 1200	1650
1350 × 1350	1800
1500 × 1050	
1500 × 1500	
1800 × 1200	
1800 × 1800	

Refer to page 5 of GGF 9, *Glass bending*, which includes a glossary of curve types.

Glass treatments

Mechanical finishing to glass Sand-blasting and etching is described in section 5.3 under 'Special glass products', p. 88.

Edge treatments There are several edge treatments to glass which should be considered for elements that are:

- Freestanding, including screens, partitions, divisions on desks, shelving, frameless doors, table or counter tops and walling details where exposed edges could be: arrised, flat, round, full round, thumb or bullnosed.
- Framed in doors, windows, screens or walling but need additional delineation through being: bevelled, flat edged or mitred.

Surface treatments Surface treatments to glass for signage and decoration include:

Cutting and engraving The quality of cut shapes and patterns is only effective if the glass is lit from the side. Glass can also be hand chipped to give an overall patterning that has a greater surface brilliance than grinding or sand-blasting.

Acid etching Glass is usually wax coated and then the design is drawn through the coating with a pointed tool. Hydrofluoric acid is then applied and etches the exposed surface of the glass. Different tones can be achieved if the strength of the acid is altered, low strength mixtures (usually moderated by alkalis) give whiter finishes as the surface topography is cruder and scatters more light.

Silvering See sections on Mirrors, pages 92–93.

Fire hazards of glass texture

Intensification of radiation through focusing capabilities of textured window glass. For this report from the Fire Research Station in 1982, Barbara Goldstone (HMSO) investigated the source of some fires in households. A fashion for *Georgian style* houses with bullion glass panes, in conjunction with the use of furnishing fabrics with a low melting point and high flammability, provided the right ingredients for fire, particularly of curtain fabrics behind the glass. Flemish rippled glass was also found to have produced burn marks, leaving tracks in a carpet. Advice is given in this report to use imitation bull's-eyes rather than original ones. One method suggested to overcome the problem was to paint the glass with clear nail varnish which reduced the intensity of focus by over 50%.

5.9 Glazing compounds, sealants and storage

When glass is glazed it should be set into rebates which are at least 8 mm deep unless there are panes under 0.1 m^2 which can then be housed in 0.6 mm rebates.

Linseed putty

The most common glazing compound is linseed oil putty complying with BS 544. It relies on being absorbed to some degree by the substrate and, although this is possible if hardwood or softwood surfaces are primed to BS 2521 or BS 2523, some dense hardwoods such as teak are unable to absorb the putty sufficiently. Coatings which seal timber surfaces completely are unsuitable.

The main constituent, linseed oil, comprises glycerides of oleic and other unsaturated acids and is classed as a drying oil and hardens easily by reacting with oxygen in the atmosphere to form a solid film (oxidative polymerization). Traditional putty is then made by mixing with *Whiting*, pure white lime and water (usually chalk). Old recipes used red and white lead to make hard putty, and lead and salad oil to make soft putty. For bedding glass thermoplastic putty often contained tallow. Putty should be protected by painting within one to two weeks. Special putty should be used for steel (except for stainless steel).

As the solid particles are suspended in the oil medium, there will be a settling of the material over time and the oil may have to be mixed back into the mixture.

Flexible compounds

This category covers a range of products from compounds which can be combined to form a rubbery phase. They have advantages by being high strength materials; useful for glazing units which have to tolerate being moved after manufacture. They are particularly useful for situations where glazing has to accommodate a greater range of movement, for example, in metal or plastic frames, or where solar control glass is used. Frames that are not painted but treated with stains show a greater range of movement, and these finishes may be incompatible with traditional puttying systems. These flexible compounds may not need timber to be primed before application but this should be checked with the manufacturer.

The non-setting compounds work by forming a protective surface skin through oxidative polymerization, leaving the body of the compound in an unset phase. These compounds can be used for insulating glass units.

Sealants

These are usually gun-applied materials which respond to a great deal of movement. They often take some time to cure. For example, the polysulphides may take between 14 and 21 days to cure and curing times of up to four months are not unknown. Silicones are rubbers which will cure between five and seven days. Some polyurethane sealants may need primers to adhere to glass or metal and they will cure between five and ten days. Some sealants are classified as *solvent release types* with butyl or acrylic bases. They change their state, remaining soft but become sticky under ultraviolet light, attracting dirt. Two-part polysulphide sealants can be tough and flexible, and will bond well to most materials.

Timber is likely to need priming before the use of these particular glazing compounds and manufacturers' advice should always be sought with regard to compatibility of primer.

Pre-formed strip materials The most common are strip materials in butyl or polyisobutene polymers, or extruded solid sections made from synthetic rubbers or PVC.

Compression gaskets These can be structural gaskets which come as a set of components that lock together putting a compression force on frame and glass. They form a tight seal and improve rigidity. The material most commonly used is a vulcanized polychloroprene or other synthetic type of rubber. Non-structural gaskets will be used as seals.

Preparation of rebates

Apart from the specific points raised in the sections above, all softwood timber should be primed, and absorbent hardwoods should be sealed. Galvanized and stainless steels will not need primers although other metals such as aluminium or bronze may need protection from particular sealants. Concrete and stone will need sealants and should be protected from possible staining.

Storage

If glass is stacked, and water allowed to penetrate between sheets, it can leach out soda from the body of the glass, transform into an alkaline liquid which can then etch the glass, and degrade or discolour it. This can happen within weeks rather than months. Early glazing should be avoided if cementitious material is still being used in close proximity on site.

5.10 Glass as a medium for performance change

Variable transmission windows

There is now a great development in the engineering of *smart glass* which can respond to external influences. These windows have the capability to change from clear to opaque transmission under the effect of visible, ultraviolet and near infrared radiation, and so will react to high solar gain. They achieve this through mechanisms which are photochromic, thermochromic or electrochromic in nature, changing the window transmission from clear to opaque, blocking solar gain and providing shading.

Thermochromic Thermochromic windows change directly with the effect of solar radiation, becoming opaque when activated by increases in heat. They rely on the properties of a stable gel formed of organic polymers which have a chain thickness less than the wavelength of light. Under the effect of solar radiation larger molecular chains are formed which start to refract the light and give a cloudy appearance. A gel thickness of only 0.3 mm will reduce transmitted light to approximately 39% which will have a diffused quality. Activation of the gel starts at temperatures of about 23 °C, and the only disadvantage can be a patchy response if the glazed area has uneven exposure to sunlight. The temperature range can also vary from 25–29 °C in direct sunlight to 27–30 °C with no sunlight.[2]

A change in the gel is also affected by the room condition as well as the solar radiation. A cool room will still allow the gel to remain clear and gain the benefit of solar radiation, a warm room will cause changes in the gel. This does have repercussions in terms of control and there is an argument for making windows with part control of glazed area so that vision can be maintained if needed.

Photochromic Photochromic glasses undergo a reversible change under sunlight using the properties of silver halides with Cu^+ as a minor component. They are expensive and use is confined to small areas of glass such as optical lenses.

Electrochromic If a small voltage is applied to a selective material in these glasses, the material changes colour to give reduced transmission and shading. The switch can be controlled to operate when solar gain reaches a particular level and shading is needed.

Window performance is vital in term of the energy performance of buildings and for optimum performance windows should have a U value of under 1.3 W/m^2 K, otherwise more heat can be lost than admitted. As the technology for variable transmission is relatively new, there may be insufficient information available on material durability or compatibiity. There should also be a good understanding of the performance needed from a window so that control can be optimized to make use of solar gain and shading achieved at the right times. When solar gain is wanted in winter conditions these windows will switch to shading when it is not needed and so over-riding mechanisms need to be installed for seasonal adjustment.

They can be controlled either by manual switching or by computer controlled switching activated by instrumentation operating on lighting levels or thermal sensors. They can help with the control of glare and limit solar gains if undesirable, which will then reduce the building's cooling load if appropriate. They give an additional layering of control which ironically enough has been enjoyed by generations of people in the Far East who will switch from open frames, giving sunlight and fresh air, to paper screens giving light without glare. For extremes in weather and at night, shutters are pulled across. This whole system of control is highly effective. However, it lacks the more discrete comfort expectations of today which seek to minimize draughts, but it is more effective in one respect by not having the external glazing acting as a black body radiating heat to the night sky.

Glass as a variable element

Glazing is the most variable element in the building facade and the scope for changing its properties is immense. *Fenestration 2000* by David Button and Reg Dunning, published by Pilkingtons, points the way forward in terms of the following six points:

Comfort How building design and investment will respond to an ageing population, and the demand by occupants for more control over their immediate environment.

Communications How the growth of information technology will create a demand for *intelligent* buildings with built-in services for office automation and communications equipment.

Visual amenity Why the visual and decorative functions of fenestration might become more prominent as building occupants demand more contact with the world outside.

Security The key factors now influencing the development of fenestration products to cope with the threats from electronic eavesdropping, intruders, terrorists and physical accidents.

Energy How long term prospects for energy and efficiency will affect the building industry and fenestration.

Materials Factors that are holding up the search for and acceptance of new materials for the building industry; and how in future glass might be seen as a high performance composite product incorporating electronic, holographic or other materials. The material commonly known as *glass* is still likely to use the existing sodium silicate technology, but with more composite constructions and additions that can modify the performance of glass and fundamentally change its properties.

As glass is a ceramic, it is essentially a stable and long life material. Developments this century have seen advances in the control of opacity and many new uses of glass, as a vehicle for insulation or as a constituent of new composites. As the making of glass improves, the material can be changed and the finish controlled to make glass that hardly requires cleaning. There are also glasses which can be engineered phased alloys with properties and performance equivalent to stone cladding materials, certainly having equivalent strength and opacity.

Transparent insulation

Systems have been developed in the last few years which construct a sandwich of glass layers enclosing insulating materials. The term *transparency* is not strictly correct as the resultant product is translucent and will appear opaque. There are various types of insulant material which will distinguish the different types being produced. The advantage is in the provision of daylighting without glare, which usually results in the use of fixed shading systems which will reduce daylight factors and yet retain the useful solar gains.

To be effective, the insulation material is honeycombed, giving a high degree of light transmission but low thermal conductivity. The material should be a good absorber of infrared light with the insulation cells of sufficient depth to minimize any escape of infrared light.

Early insulation materials developed by Okalux used polymethyl methacrylate (PMMA). This is largely being replaced by glass capillaries which meet standards for fire resistance. Foam materials are also available but are not as effective as the honeycomb structure.[3]

Aerogel insulation Aerogels are materials with an air content of about 90%. They are becoming more common as insulants in combination with glass to give fixed transmission and thermal performance. Their use in solar panels and as CFC-free insulants is due to their unique properties of scattering light but giving improved illumination. They allow the passage of 50–80% visible light making them ideal for use as solar collectors in front of walls. The evacuated aerogel has a U value of 0.65 $W/m^2 K$. The solar transmission of the aerogel is about 70% and the enclosing glass 80%.

The other major use of transparent insulation involves the cladding of masonry with a glass skin that has an air gap, transparent insulation and then a masonry wall. Solar energy passes through the glass, air gap and the insulation diffusing through the masonry wall which acts as a longer-term storage for heat. Energy is trapped and cannot pass back through the outer wall. This system, used experimentally in Germany, is showing good results. The space between the glass and the insulation can become extremely hot and the building fabric can overheat to an extent where blind systems may have to be introduced within the air gap. This gives an extra level of control which is difficult to maintain in a remote situation. Overheating can also cause problems with the framing subassemblies which can move and deteriorate, losing their weather protection. The differential expansion of glass which will happen at different rates to framing assemblies is magnified under these conditions.

Future developments may use thermochromic or electrochromic control systems which would eliminate mechanical blind systems which are difficult to maintain. Holographic systems being developed could also give the same degree of variable shading control.

There are direct benefits in retrofitting facades in existing buildings and cladding with glass skins. This allows the original masonry wall behind to become a longer term storage unit although the walling mass should be calculated to make sure it really is that effective. This strategy is very effective with glass block walling systems. There are negligible heat losses through transparent insulation systems. The weakest links are the framing systems which can be a source of cold bridging. Framing technologies are not keeping pace with the development in more efficient thermal glazing systems. Demonstration projects conceived and monitored by the Fraunhofer Institute for Solar Energy Systems show it is possible to contribute between 100 and 200 kWh/m^2 to the heating demand of a building.[4]

Another use of transparent insulation is as a cover to plate collectors. This makes the collection of solar energy more viable in northern and central European climates. Experiments in cladding water tanks show the possibility of keeping water storage at a preheat temperature which would save energy in hot water systems, as less energy would be needed to bring the water up to the required temperature.

Polycarbonate honeycomb insulation On cold days, the glazing provides effective insulation, giving a U value of 0.72 $W/m^2 K$. This can be improved by using a roller blind giving a U value of 0.46 $W/m^2 K$. This insulation can also give sound reductions of 50 dB.[5]

Another system uses capillary slabs sandwiched between glass. Light transmittance can be varied between 0 and 72%. There is a recommendation for glazed roofs to have lighting transmission values of between 40 and 60%. Sound level reductions of 51 dB can be achieved with this system. Unit sizes are limited and currently should not exceed 800 × 3000 mm. Roof constructions will normally use a sandwich construction with 7 mm wired glass, a capillary structured glass slab and rough glass of 7 mm. Alternatively, laminated glass construction will use 8 mm glass with fibre screens either side of the capillary slab, and a float glass of 6 mm. U values of 1.14 W/m^2 K can be achieved if the maximum thickness of the capillary slab is used (40 mm). Slab thicknesses can otherwise be 8, 12, 16 or 24 mm.

As a further option to this sandwich construction, clear plastic tubes can be arranged as a honeycomb then coated with a transparent film or opaque glass fibre tissue.

Holographic systems

Using active materials in the glass, it is possible to produce holographic projections which can be reflected to show instrumentation (currently used on military aircraft), or could in the future show any kind of static or moving image, with displays built up from pixels of information.

Reflective solar glasses

A new sandwich construction incorporating chromatized louvres which can bounce daylight into a building has been developed.[6] With regard to solar gain it can give a reflected transmission of 19% and a direct transmission of 6% in summer; and in winter a reflected transmission of 38% and a direct transmission of 7%. In summer the direct light transmission can be between 5 and 10% and in winter between 45 and 60%. Throughout the year diffused light transmission will vary between 45 and 60%. This gives light without glare, with improved performance in winter when it is most needed.

For these glazed sandwich constructions to be effective, particular attention must be paid to their detailing. Framing systems can be a source of cold bridging and air infiltration which would reduce the effectiveness of a total cladding system. There is a need to evaluate the whole construction and the total thermal performance rather than predicting high performance on the basis of an individual panel.

Insulation products

Glass fibre Glass fibre has been a standard form of insulation and is also used in mat form in sheet composites and as chopped fibres for reinforcement in polyester as glass reinforced polymers (GRP), with cement as glass reinforced composites (GRC), and in concrete. The fibres are made by pushing molten glass through holes in a spinnaret. They are inherently strong but any defects, such as notches or scratches, can weaken them. The fibres are coated for protection to prevent fracture through any surface imperfections before use, usually with a polyester resin. Glass fibre is generally used in a bulk form in the building industry often as mats which can be 'bonded or unbonded, woven, felted, carded or needled, and can be supplied in rolls or bats' according to BS 3533. *Quilted* as a term usually refers to material that is faced on both sides with a building paper or is totally enclosed.

Mineral fibre Glass fibre or glass wool is still defined as mineral fibre and mineral fibre is a general term used for all non-metallic inorganic fibres made from molten glass, rock or slag. For main definitions of insulation refer to BS 3533 1981. Ceramic fibre is mostly silica, SiO_2.

These fibre-based insulation products are often made into composite boards. For example, the Mataki HT roofing system has a system where polyester resin based bitumen and felt is bonded to rock-wool. Thick skin coatings have also been developed using rock-wool, for example, for external wall refurbishment systems.

Asbestos Asbestos is included in the ceramics section because the minerals fall into the category of silicates, usually magnesium iron calcium silicates, with or without aluminium. These silicates are unusual as although chrysotile (a common mineral in the asbestos group) is made up of sheet silicate structures the sheets roll up to form tubes, then having the appearance of long flexible fibres. The linear fibres are separated from each other by cations (atomic particles or molecules which carry a positive charge) and as like charges repel each other the structures remain as separate strands.

Glass and other mineral fibres mimic the appearance of these natural fibres but are usually made artificially by a reduction of the raw materials to a molton solution. This is then blown through a spinnaret into fine strands which solidify on cooling.

The general formula for magnesium iron silicates is:

$$4[(Fe\ MgAl)_7\ (SiAl)_8\ O_2\ 2(OH)_2]$$

5.11 Health

Exposure to mineral wools, Guidance Note EH 46, *Health & Safety Executive* requires a one-piece overall, gloves, eye and respiratory protection, and minimal cutting and handling on site.

Man-made mineral fibres cause irritation to skin and eyes and to the upper respiratory tract. European studies show an increase in lung cancers in workers involved with production over a 30-year period. There are recommended exposure values for wool dust in the atmosphere. Dust production is prevented by not cutting to size and keeping the material under a protective covering or by using materials that incorporate oils or binders. Handling in restricted unventilated areas (such as lofts) is hazardous and exceeds the recommended safety levels. Overhead handling increases personal risk. The size of fibres is the critical factor which increases the carcinogenic properties, not the mineral composition, and as an approximate guide the most dangerous are those which have lengths greater than 8 microns and a diameter of less than 0.25 microns. This size of fibre can penetrate the lungs contributing to their permanent damage through the formation of tumours.

Reference should be made to the approved code of practice published by the Health and Safety Commission, *Control of substances hazardous to health regulations, 1988*. There is also the document: *The control of substances hazardous to health in the construction industry*, published by HMSO. *Hazardous building materials*, edited S R Curwell and C G March, published by Spon, 1986, is also a useful guide to the possible health risks for all materials.

Asbestos removal

The removal of asbestos materials has substantial legislation separate from the *Control of substances hazardous to health* documents published by the Health and Safety Commission. The type of asbestos has to be identified first as this will give the particular mineral which will then have certain fibre characteristics. A complete list of all relevant publications is given in the *Asbestos subject catalogue*, a free leaflet from any office of HMSO. Removal of blue asbestos will immediately exceed any control limits or action levels and there should be enclosures provided to isolate areas that are being worked on, Workers should be properly clad and the air quality of sites should be monitored while work is in progress. No safe levels have been established for working with this material, and length of exposure is not a factor, i.e. risk can arise from relatively short-term exposure to the fibres. Respirators should be worn if the limits set out in guidance note EH 35 *Probable asbestos dust concentrations at construction processes* are exceeded.

References

1 J McColm 1983 *Ceramic science for materials technologists* Leonard Hill: pp 194.
2 E Boy Meinhardt 1988 TALD: A temperature-controlled variable transparent glass. *Building Research Practice* **16** (4) Spon.
3 A Goetzberger 1989 *Transparent insulation: A new solar energy component*.
4 W Stahl, E Bollin, J Schmid, J Vahldiek, J Voss and A Wagner 1989 *Wall heating with transparent insulation: Results from realised demonstration projects* CEC Kluwer Academic. Paper presented at the Second European Conference on Architecture.
5 H F O Muller and N Kaiser 1989 *Development of controllable facade elements with transparent insulation material*. Workshop proceedings of Transparent Insulation Technology.
6 'Oka Solar' produced by Kapillarglas, 8772 Marktheidenfield-Altfeld, Germany.

III METALS

6 Introduction

6.1 Origin of metals

Metals used in the building industry are far removed from their origins as sometimes complex mineral substances. As materials we have a familiarity with metal products but very little knowledge of how these relate to their mineral origin. Most metals used in the building industry have undergone extensive processing from basic ores, which can still contain between 10 to 30% waste material in the form of earth, sand clay and rocks.

Communition techniques are used to crush and release higher grades of ore. Further separation techniques are carried out using density differences, magnetic separation and even frothy liquids which attach themselves electrostatically to particles to separate copper, zinc, tin and lead ore from unwanted earthy matter. The silica content in rocks is difficult to remove because of the high temperatures involved, and fluxes are used to decrease the temperatures needed. Further thermal or electrical separation is used (particularly for aluminium) to extract the refined metals we eventually use. Even these metals used are not always *pure* in the sense of being composed of single elements, except for copper, lead and zinc in roofing. In order for them to be sufficiently stiff for their purpose they are often combined with other elements and are generally alloyed with other metals or substances. Their use and specification is related to the control which has to be maintained not only of their constituents but also the method of processing, as heat and cooling treatments are used to control the formation of crystal structure and consequently a wide range of properties.

Before ores reach the early stages of metal processing they are won from the earth by mining, which involves the mass movement of raw material in order to extract ore. The scale of mining is difficult to appreciate, but opencast mining will affect vast areas of land, and underground mining will develop hidden structures that dwarf our concepts of three-dimensional structures above ground. As sources of ore diminish due to working, the percentage of ore found in the ground will drop and extraction becomes deeper and more difficult. There comes a point when the amount of ore extracted is uneconomic, due to the amount of men, equipment and time needed. At this point the ore is known to have a *cut off grade*, and workings will move to other areas or the winning of that particular ore will be frozen until there is a rise in the market price of the metal, to ensure some profitability in extraction.

As an example, a copper ore body at the turn of the century used to contain between 5 and 8% copper, in the 1930s this dropped to about 2%, and copper is now being mined with ores containing less than 1% copper, with cut off grades in the USA of around 0.4%. The price of copper has risen steeply in response. As techniques in mining improve, previously inaccessible ores become available, but generally reserves are being depleted with projected usage anticipating the life expectancy of reserves to be in 1982 as shown in Table 6.1.

If projections prove correct then most metals will not be available from primary sources after the first ten years of the twenty-first century and will be used within one generation.

Territorial disagreements are more likely to be centred on the protection of access to resources rather than

Table 6.1 Life expectancy of ore reserves

Metal	Life expectancy of reserves (year)	
	1982	1991*
Lead	29	20
Tin	31	22
Copper	35	24
Aluminium (bauxite)	94	85
Zinc	21	12

* Extrapolated from 1982 usage

fundamental difference of nation states with regard to land ownership or political or religious disagreements. The Falklands War was prompted by an interest in oil reserves below the sea floor east of Argentina as well as the protection of access to possibly rich mineral deposits in Antarctica. The recent Gulf Crisis has been more open in the obvious protection of oil supplies for the West.

One quarter of the world's population uses three-quarters of the world's non-fuel minerals. As the processing of these minerals is energy intensive there is an important relationship with energy consumption and future ability to process on a large scale as energy resources diminish. In order to process materials it becomes increasingly important to conserve energy in other areas and this puts pressure on two major areas. One is the design of energy-efficient buildings with complete conversion of all existing stock so that energy is not wasted; and the second is that the decision making choice of materials has one over-riding parameter, which is its energy efficiency rating in terms of primary as well as product processing. The movement of raw materials in such bulk is also energy intensive, and consequently there are four main energy demands in the production of metals:

- Mining and ore-dressing at source
- Transportation
- Extraction using heat or electrical energies
- Processing to finished product

This data comes from the *Global 2000* report to the president of the USA, published in 1982 by Pelican Books. It also predicts the mining of refuse dumps in the future for material recovery. In the twentieth century it is common to reach cut off grades, and emphasis is then transferred to the recovery of metals already in circulation, and the organized collection and refining of scrap. As mining progressively adds more metals to an active global reserve, the amount of metals in circulation rises. In the USA at least 50% of copper demand is met from scrap, 50% for lead, 31% for iron and 12% for zinc. Aluminium is also an important recyclable metal. It becomes important to maintain the quality of metals in circulation and to some extent this is hampered by the complex alloying of metals which makes refinement back to a pure elemental state nearly impossible. It becomes increasingly difficult to maintain a purity in the recycling of metals, particularly for steel. Recycled metals can be progressively downgraded and less fit for components which demand a particular specification.

The United Nations now, where possible, include figures in their statistical yearbook which show tonnage of not only primary but also secondary processing of materials. For example, in the UK in 1987 294.4 million tonnes of aluminium was produced through primary processing, but an additional 116.7 million tonnes was produced from secondary processing. The secondary processing accounts for some 30% of aluminium supplied. Lead has always had a high recirculatory value and in 1987 in the UK, although 373.6 million tonnes was produced from primary processing, 668.4 million tonnes came from secondary processing.[1] The production of steel in primary processing in the UK increased from 6316 million tonnes in 1980 to 17,414 tonnes in 1987. There are no equivalent figures for secondary processing but this shows the level of production increasing almost three fold. These figures may help reflect trends in building. For example, cement production in the UK actually fell from 15,916 tonnes in 1978 to 14,311 tonnes in 1987. For a full picture, imports should be taken into account, but in the last few years there has been a boom in building with steel and the demand has been led by commercial needs for fast track building which could be satisfied by this material.

Natural ores are found in combination with oxygen, silicon and often sulphides, and are rock-like in their character. They are stable compounds and any further processing or extraction to a pure elemental form puts most of them in an artificial state of high energy. The initial brightness of most metals soon dulls as they combine, usually with oxygen in the atmosphere, before more complex reactions initiate a return to a compound with a lower form of free energy. This is termed *corrosion*. Control of corrosion is one of the most important concerns in the building industry, not only with regard to safety of components but also with the maintenance of metals in circulation as an important resource.

Principles of alloying

Most metalic materials in general use are alloys since this allows their properties to be manipulated for the most appropriate performance in use. Although alloys are mixtures of identifiable individual metals their combination can precipitate quite complex intermetallic compounds. In a solution of metals, after cooling and solidification, it may be possible to identify the original pure metal constituents, or there may be new combinations with two metals yielding a new solid solution, or one identifiable metal may be suspended in a new solution. All of these variants are individually identifiable to the metallurgist and they are called *phases*. The formation of different but identifiable phases is controlled by the temperature the metals are heated to and their rate of cooling as well as the relative proportions of the original constituents used in the mixture. Atoms diffuse through metals at different rates at different temperatures and this rate of movement is arrested when the cooling process starts. Consequently the final phase structure is determined by the rate of cooling. This is why subsequent heat treatments to metals have to be carefully controlled. Phase structures can be altered and this will affect the strength and other properties of the metal explaining why welding is such a critical process and has to follow precise standards.

Trade associations

For practical help each metal has its own trade association, most of which provide quite comprehensive literature. As most of the trade practices, especially for copper, lead and zinc are craft-based systems of working, the literature is very clear, practical and well illustrated.

6.2 Corrosion susceptibility and prevention

As most of the metals used in the building industry are alloys, Table 6.2 is a better indicator than the electrochemical series which deals with pure metals.

As different metals and alloys form oxidized coatings this gives a protective mechanism and even in water-free oxygen will contribute to these types of passive layers. However, sometimes in crevices there is oxygen depletion and this passive layer cannot form. These areas are known as *de-aeration cells* and can lower the electrode potential and increase the risk of corrosion.

Table 6.2 Relative corrosion susceptibility

Noble metals
Platinum
Gold
Stainless steel 18/8/3 Mo type 316 (passive)
Stainless steel 18/8 type 034 (passive)
Titanium (and its alloys)
Monel
Copper–nickel alloys
Copper
Brasses
Tin
Lead
Stainless steel 18/8/3 Mo type 316 (active)
Stainless steel 18/18 type 304 (active)
Lead–tin solder 50/50
Stainless steel 13 type 410 (active)
Cast iron
Mild steel
Cadmium
Aluminium (and its alloys)
Zinc (and its alloys)
Magnesium (and its alloys)
Base metals

Source: Guide to engineered materials 1986 ASM, Ohio, which shows the galvanic relationship of metals in water

The nature of atmospheric corrosion and electrochemical corrosion

There is a natural tendency for many metals to corrode. This can be understood in terms of free combination with elements in the air to form oxides and, less commonly, sulphides. Some of these compounds formed on the surface of metals have sufficient integrity to protect the metal from further corrosion. These films can be improved and controlled by the alloying elements in metals. Aluminium, lead, stainless steel and copper also produce environmentally stable films, but ordinary iron and steel produces corrosion products that are discontinuous, leaving the underlying metal exposed to further change. Corrosion on the surface of ordinary steels can usually be seen as a series of isolated spots that increase and deepen to characteristic pits. There are two major equations explaining how iron dissociates.

(1) $Fe \rightarrow Fe^{2+} + 2e^-$

iron → iron ions (positive) + free electrons (negative)

(2) $\frac{1}{2}O_2 + H_2O + 2e^- \rightarrow 2OH^-$

oxygen + water + free electrons from (1) → hydroxyl ions

The products of both of these reactions combine:

$Fe^{2+} + 2OH^- \rightarrow Fe(OH)_2$ (ferrous hydroxide)

This is not the final product as it oxidizes further to form hydrated ferric oxide:

$Fe(OH)_2 + \frac{1}{2}O_2 + \frac{1}{2}H_2 \rightarrow Fe(OH)_3$ (rust)

which breaks down to a hydrated oxide FeO.OH. Rust is finally a mixture of Fe_2O_3 (ferric oxide) with Fe_3O_4 (magnetite) and the oxide FeO.OH.

The process of electrochemical corrosion

Every metal has a potential to corrode and different metals will have differing potentials which can be measured relative to a known standard. One standard commonly used is hydrogen which is given a datum value of 0.00. This rates a metal's ability to revert to its more natural state. Precious metals are valued for their durability because they have a low potential to change, i.e. they are found in their elemental state as pure silver and gold. They need no refining and are known as *noble metals* with a *positive* measured potential. Other metals found in ores in the earth's crust that need extraction have *negative* potential. All these base metals have a rank order and the more complex their ore chemistry and the more difficult they are to extract, the more likely they are to revert or corrode, and the more

Table 6.3 The order of potential of some common metals

Characteristic	Metal	E(V)
Noble metals	Gold	+1.42
	Silver	+0.82
	Copper	+0.34
Reference	Hydrogen	0.00
	Lead	−0.13
	Iron	−0.44
	Chromium	−0.70
	Zinc	−0.76
Base metals	Aluminium	−1.66

negative their electrode potential. Common metals are given in Table 6.3 in their order of potential.

Bi-metallic corrosion If two differing metals or alloys are adjacent, and if there is a medium which allows for the passage of electrons (and hence a current to be carried), the difference in potential can generate the formation of ions, and the metals form a corrosion couple known as a *galvanic cell*. Electrons travel from the more negative metal (*anode*) to the more positive metal (*cathode*). This can be explained in a very broad equation:

$$M \rightarrow M^{n+} + e^{n-}$$

Metal → free metal ions + free electrons

Base metals have this tendency to dissociate. Noble metals have the same reaction but in the opposite direction.

$$M^{n+} + e^{n-} \rightarrow M$$

free metal ions + electrons → metal

Rates of corrosion Rates of corrosion are related directly to how fast these electrochemical reactions can proceed, and they increase with higher temperatures, greater humidity and usually a steady oxygen supply. However, corrosion can occur more rapidly in areas that have a limited oxygen supply, i.e. in concealed corners or under dirt deposits, and these are known as *differential aeration cells*. In practical terms, real rates of corrosion are affected by climatic conditions. Dry rural areas show a slow rate of corrosion, and in the corrosion of mild steel this would be about 0.25 microns per year in the Sudan, increasing to 80 microns per year (or the equivalent of four paint films) in marine and industrial atmospheres. In tropical marine atmospheres, such as Nigeria, this can increase to 620 microns (over half a millimetre) which is recognizable and substantial damage to a metal surface. These figures are relevant, bearing in mind that we can often emulate tropical marine environments in particular situations (e.g. swimming pools). This poses a serious problem in ensuring that our metals are kept in a passive state and free from corrosion.

Prevention of electrochemical corrosion using inhibiting systems

Electrode potentials in metals have implications with regard to common metals used in the building industry. As a great variety of metals are used in this industry, there is a potential for galvanic reactions to take place. In the table for electrode potentials a value is given for iron, but steels have a slightly more negative value due to their alloying elements, and a greater potential to corrode. There can be differences between alloys of the same metal, which is why specifiers should try and ensure that compatible alloys are used adjacent to each other. Exposed finishes, which include internal and external cladding systems and their fixing details, should be looked at carefully as thin section metals are more vulnerable. A slow rate of corrosion may be acceptable on some heavy structures, and the corrosion acceptability of a detail will then determine its method of preparation and protection. Although some metals (e.g. aluminium) have theoretical high differences in potential from steel, their tenacious oxide coatings or their finishes may inhibit the free passage of ions. Proof of differences in potential can easily be measured and catered for. If some junctions are encouraging corrosion by the abutment of dissimilar metals, gaskets and separating mediums should be specified to achieve *electrical isolation*.

As electrochemical corrosion is based on electrical activity in one direction, a current reversal induced artificially can stabilize the chemical situation and prevent corrosion. This is a specialized area but some coating systems are now being designed to act as an electrolyte to carry a weak charge to protect reinforcement in concrete. The complexity of the system requires some management to be built in to such an installation ensuring that potentials between different areas are monitored. They can only be used to protect elements in the ground or in water at present. They are known as *impressed current systems*.

Corrosion inhibitors are commonly used in pipework systems (such as central heating) and work by removing oxygen from circulating fluids, or by producing a passive environment which is normally alkaline. Those inhibitors which produce this passive environment are often used as first applications to steel, prior to the application of the full coating system.

The metals most common in the building industry which professionals need some knowledge of are given in Table 6.4.

It is rare for pure metals to be used in the building industry and, generally speaking, the metalic materials used are all *alloys*, i.e. their properties and behaviour have been

Table 6.4 Common metals and their ores

Metal	Ore	Chemical name	Formula
Aluminium	Bauxite	Hydrated aluminium oxide	AlO.3OH
Copper	Copper pyrites	Copper iron sulphide	CuFeS
Iron	Haematite	Iron oxide	FeO
Lead	Galena	Lead sulphide	PbS
Nickel	Pentlandite	Nickel iron sulphide	$NiFeS_2$
Zinc	Zinc blende	Zinc blende	ZnS
	Carmine	Zinc carbonate	$ZnCO_3$

modified by the addition of other elements, which can be metallic, non-metallic or a combination of both.

6.3 Microstructure of metals

Most metals used in the building industry, chromium, manganese, iron, copper, nickel and zinc are known as the *transition metals* and form one continuous block in the periodic table from Group IIIB to IIB. They have high melting points and high densities, which increase with their atomic number. Aluminium stands outside this block and is in Group III of the periodic table with a much lower density and melting point.

All metals have properties which are similar. They can be deformed easily due to their structure, they are good thermal conductors and can conduct electricity, although the performance of conduction varies greatly and is best in copper, silver and gold.

Metals are crystalline materials. The atoms are packed regularly to form crystal shapes which can be seen as irregular grains. The size of these grains depends on how quickly the metal has been cooled. Slow cooling produces large grains which can be seen with the naked eye, but fast cooling produces small grains which can only be seen under the microscope.

At an atomic level, metals have a regular packed structure which can be hexagonal close packed (zinc and magnesium), cubic close packed (aluminium, copper, lead, silver, gold and platinum) and body centred cubic structure (iron and manganese–alkali metals). These geometric structures repeat in a crystal which will be composed of millions of atoms. Metals have a commonality in their method of bonding which is known simply as metallic bonding.

Instead of the outermost electrons in atoms being tightly bound, they move freely in what is often described as a sea of electrons. This does not allow for the clear passage of light, which would happen in materials with a tightly bound structure, so metals appear opaque. Due to the movement of electrons, there is some emission of light as they fall to lower energy levels and this explains the lustre-like quality of metals. It has been possible to freeze metals at such low temperatures that the movement of ions is inhibited and the resulting material is then glass-like. Only minute amounts of glass-like metals have been made and they cannot stay stable because of the very low temperatures needed to maintain them in this state.

As the bonding is not rigid, this means that layers of atoms can slide relative to each other, and explains the malleability of metals and their ability to be formed into complex shapes by mechanical pressure. Sometimes this is not an advantage in structural metals, and methods of processing are used to prevent the slippage of atoms. The boundary between the slipped and unslipped part of a crystal is known as a *dislocation line* and these dislocations can be complex. If pressure is put on a metal, dislocations can interact and start to prevent slippage. This mechanism is called *strain hardening*, and there comes a point when the metal cannot be deformed further without fracture. If more work is needed on the metal it has to be annealed, i.e. it has to be heated to a point just above its recrystallization temperature, when new grains or crystals form without dislocations and the metal can be reworked. If particles are introduced in the metal by designing an alloy to have secondary phase material, these prevent dislocations moving through the metal and stress put on the metal is effectively resisted. It does, however, mean that the metal can be so hard that it will take a much increased load but will suffer sudden fracture at high loadings.

Metallurgists are concerned with the engineering and monitoring of the microstructure so that the transformations occurring in the metal are carefully controlled. This level of control affects hardness, ductility and tensile strength, and the microstructure of a particular metallic material can be *read* under the microscope to identify the different alloys present and how the material was processed and treated. This level of investigation is used to check the manufacturing of an alloy in progress as well as to analyse failure in welds or metals in service generally.

The forming of metal to its final production profile occurs in combination with temperature controlled conditions. Working metal is a process that alters the characteristics and strength of the metal. The energy used in working is partly stored in the atomic lattice of the metal, it changes the crystal structure and strengthens the material. In hot rolling this gives added toughness, but in cold rolling the deformation can make the materials so hard that it is more liable to fracture in a brittle manner under stress.

Cold working is the mechanical deformation of a metal below its recrystallization temperature. This relates to metals that might be rolled, hammered or even drawn to produce tubes and wire. A great deal of energy is released, about 90% being released as heat, the other 10% is stored in the atomic lattice. Metals that have been subjected to cold working have elongated crystal grains, deformed in the direction of applied stress. Cold worked metals have a greater surface accuracy in machining but have the disadvantage of being subject to higher rates of corrosion. Hot worked metals have mechanical deformation carried out above their recrystallization temperature but the metal is weaker and the surface accuracy is not so good.

6.4 Testing of metals

All metals need evaluation for strength and performance, and quality control in the processing of metals is achieved by the continual testing of metal as it is produced. The testing methods may produce results that are not obvious to the naked eye and specimens have to be examined by metallurgists under the microscope to detect changes in the microstructure of the metal.

Different metals have different known values of strength and toughness, relative to the alloys made and the heat treatments used. Testing methods are important to determine values and to ensure that standards are adhered to. Methods determine the yield strength, ductility and stiffness through tensile testing, toughness through impact testing, and hardness through resistance to penetration of the surface by a hard object.

Tensile testing

Tensile testing involves the application of load onto a specimen and physically stretching it under an increasing load, measuring the extension of the metal and the reduction of the cross-sectional area until fracture occurs, often referred to as *necking*. Testing can be carried out on solid rods which can be round or flat, or tubular rods. Tensile testing records the limit at which a metal behaves elastically, i.e. once the load is released it can return to its original state. Once the elastic limit is passed the metal will flow plastically and fail under a lower applied loading.

The tensile strength is the stress corresponding to the maximum force applied.

$$\text{Stress} = \frac{\text{load in Newtons}}{\text{cross-sectional area of specimen}}$$

Strain is the degree of extension of the metal expressed as the actual extension over the original length of the material.

$$\text{Strain} = \frac{\text{increase in length}}{\text{original length}}$$

The relationship between stress and strain is a measure of the elasticity of the material. If a load is applied to a material, and that material makes a recovery, i.e. returns to its original length once the load is removed, the elastic limit of the material has not been reached. However, if the material does not recover on removal of the load, it must have exceeded its elastic limit and will be said to have deformed permanently and started to behave plastically. The elastic behaviour can be described by the relationship between the stress applied and the resultant strain, this is known as Young's modulus and is expressed mathematically as:

$$\text{Young's modulus} = \frac{\text{stress}}{\text{strain}}$$

Young's modulus of elasticity is the ratio of stress to strain during the elastic behaviour of the metal and is a constant in metals allowing them to be compared with each other.

This relationship allows comparisons to be made between different materials, and as the value of Young's modulus changes between materials, so then does their elastic behaviour, which is a real indicator of their stiffness. Steel has a high value of Young's modulus which is about 205 kN/mm^2 (approximately three times the value for aluminium).

In understanding fully the behaviour of steel, stress is plotted against strain as shown in Figure 6.1.

(1) The initial part of the curve OA shows a linear relationship between stress and strain. This is the *elastic* region. The slope of OA is a measure of the material

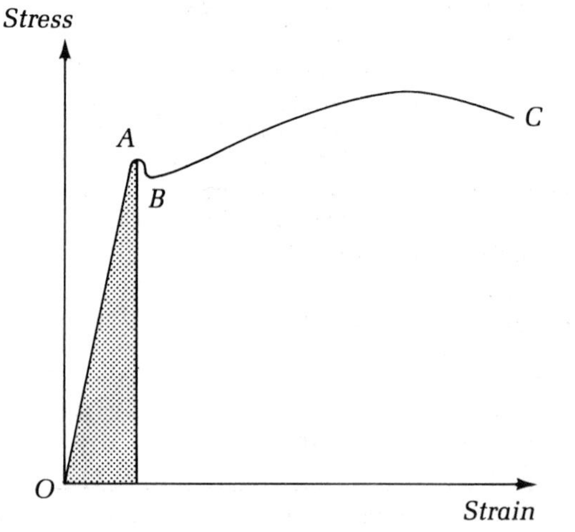

Figure 6.1 The characteristic stress–strain curve for steel. OA — elastic region, BC — plastic region, failure occurs at point C.

stiffness — the steeper the slope, the smaller the deformation for a given load — and is referred to as Young's modulus. For all steels, regardless of chemical composition or processing, the value of Young's modulus is virtually constant at about 205 kN/mm^2.

(2) Beyond point A on the stress–strain curve, further increase in load produces very large increases in strain. This is the *plastic* region. Although there is still a considerable reserve of strength, the yield stress is normally used as the failure criterion when assessing the strength of a structural member because of the very large deformations associated with plastic behaviour. Furthermore, strains within this range are associated with permanent deformations.

(3) Beyond the plastic region, further increases in strain are associated with more pronounced increases in stress. This is due to strain hardening. Eventually, the specimen will reach a maximum level of stress (point C) and fail. This is the *ultimate tensile strength* of the steel.

(4) The ability of the steel to sustain high strains prior to failure is a measure of its *ductility*. Both yield strength, and ultimate tensile strength can be increased by altering the chemical composition of the steel, particularly the carbon content. Yield strength can also be improved by mechanically working the steel, although this reduces its ductility as can be seen.

(5) An important property associated with elastic behaviour is the amount of energy stored within a stressed steel element. This is measured by the area under the linear part of the stress–strain curve. This is shown shaded. Clearly, for highly stressed conditions, such as may be found in cable structures where steels with a very high yield strength are often used, this quantity can be very large. If the cable fractures, the sudden release of energy can cause violent and dangerous whipping of the cable.

Impact testing

This is used to show how the metal would behave under mechanical shock. It is an indicator of toughness and is

Table 6.5 Typical mechanical properties of some metals and alloys

Metal or alloy	Condition	0.1% proof strength (N/mm^2)	Tensile strength (N/mm^2)	Specific strength (N/mm^2)	Young modulus (kN/mm^2)	Specific modulus (kN/mm^2)	Elongation (%)	Hardness (Brinell)
Lead	Soft sheet	—	18	1.54	16	1.37	65	4
Aluminium	Wrought and annealed	—	60	21.8	70	25.9	60	15
Duralumin	Extruded and fully heat-treated	275	430	154	71	25.4	15	115
Magnesium-6Al/1Zn	Extruded bar	170	300	167	48	26.7	10	60
Copper	Wrought and annealed	46	216	24.1	130	14.5	60	42
70/30 Brass	Annealed	85	320	37.6	100	11.7	**68**	62
	Deep-drawn	370	465	54.6			19	132
Phosphor bronze (5% tin)	Rolled and annealed	120	340	38.1	101	11.3	66	72
	Hard-rolled	650	710	79.6			5	188
Mild steel	Hot-rolled sheet	270	400	50.8	210	26.7	28	100
0.45% carbon steel	Normalised	420	665	84.7	200	25.4	27	152
	Water-quenched and tempered at 600°C	540	780	99.4			25	200
4Ni/Cr/Mo steel	Air-hardened and tempered at 300°C	**1200**	**1550**	198	**225**	28.7	12	**444**
18/8 stainless steel	Softened	185	525	66.3	220	27.8	30	170
Grey cast iron	As cast	—	300	40.5	150	20.3	0	250
Titanium (commercially pure)	Annealed sheet	370	450	100	120	26.7	30	—
Titanium alloy (4Sn/4Al/4Mo/0.5Si)	Precipitation hardened	**1200**	1390	**309**	150	**33.3**	16	—

Note: **Bold type** denotes maximum value in that property (where relevant).

Source: Table 2.1 R A Higgins 1984 *Engineering metallurgy* Vol. 1 *Applied physical metallurgy* Hodder and Stoughton

carried out by swinging a pendulum which breaks through a notched specimen. The height the pendulum reaches after swinging through is an indicator of how much energy has been absorbed by the specimen.

Hardness testing

The Brinell hardness test is the industry's standard and uses a tungsten carbide ball that is forced into the material under the test force. The diameter of the indentation is then measured.

The Brinell hardness number, H_B, is a number obtained by dividing the test force by the calculated curved surface indented by the ball.

Table 6.5 gives values of f/D^2 showing the relative hardness of different metals and alloys (f = force D = diameter), along with some other typical mechanical properties.

Vicker's hardness test (Given in BS 427:1990.) A pyramid is indented on the surface of metal by a diamond in the form of a pyramid with a square base from a known *applied force*. The depth of the indentation is evaluated by measuring the two diagonals of the indented pyramid and calculating their mean. The Vicker's hardness is a number obtained by dividing the test force by the calculated sloping surface area of the indentation made in the metal.

Fatigue testing

Metals that fail from fatigue have responded to cyclic loading. This kind of failure is more common in general engineering that uses components subject to dynamic loading. Cyclic loading progressively affects the crystal structure of metals, and failure occurs at a load far lower than the expected strength of the metal. This is why failure can be sudden and catastrophic, and is of great concern in the prediction of failure by the aerospace industry.

Heavy equipment operating in buildings can cause structures to vibrate sympathetically, and failure can be caused by fatigue. Failure occurs through fracture initiated by a small crack. The cyclic loading pattern rubs the faces of the crack edges together, they become burnished and the crack grows. Fracture occurs when the cross-sectional area is significantly reduced, explaining how failure occurs at relatively low loading.

Surface finish is important in metals that are subject to high stress and it is important that their surface is polished to remove any minor imperfections which could be crack initiators. Testing involves using cyclic loading and then determining the number of cycles at which failure occurs, and the safe operating load if a metal has to be subjected to this kind of loading.

Creep

Failure of metals due to creep is unusual and applies to metals that will deform under pressure. The most obvious example in the building industry is the behaviour of lead which can deform under its own load if fixed or detailed inadequately. Testing involves loading a specimen with a lever system and the extension of the metal can then be measured easily.

6.5 Working and processing of metals

Cold working

Cold working includes cold rolling of sheets, the drawing of solid and hollow sections (which includes the making of wire) and cold pressing. These processes generally use tensile forces to form the metal.

Spinning is also a cold metal working process and forces a circular piece of metal into a thinner section. Saucepans are commonly produced using this method. The action of cold working is also work-hardening, and energy from the processing is partly absorbed in the atomic lattice and increases strength. The surface quality of metals after cold working is good and dimensions are accurate.

The grain or crystal structure of the metal is altered showing extension in the direction of stress applied and is a clue as to how a particular metal has been processed.

Annealing Cold working can put metals under sufficient mechanical stress to increase their potential for brittle failure. After cold working has been carried out, metals may be annealed, i.e. heated at a relatively low temperature to relieve stress which allows some movement of atoms within the lattice to remove strain, or by heating at a higher temperature that enables recrystallization to take place. This usually happens between one third (for pure metals) or one half (for alloys) of the melting point temperature of the metal. After stress relieving, the metals may be worked further.

Hot working

These processes are carried out above the recrystallization temperature of the metal. Metals are softer and more malleable at these temperatures and will then recrystallize after cooling. Compressive forces are used to manipulate the metal in processing. The surface accuracy of hot worked metals is not as good as in cold working and dimensional tolerances may need to be as high as $\pm 12\%$. Hot working processes use rolling, forging or extrusion, which can also produce wire. After recrystallization the grains will show no directionality. Hot working means that less stress has to be applied to deform the metal and this may be an important factor in processing techniques.

Processing

The behaviour of metals under stress in combination with temperature control allows a great range of processing techniques.

Pressing/deep drawing This may be referred to as deep drawing and is a cold working technique which deforms the metal to a reverse shape made as an underlying mould known as a *die*. Domestic items, such as sinks and panelled materials for car bodies, are pressed in this way. The process requires the use of a ductile or malleable metal alloy which can be deformed without fracture. This means that choice of the metal used leads to use of lower-strength alloys which have to be stiffened by the three-dimensional geometric qualities of the finished artifact.

Casting Casting is a highly specialist technique which involves the flow of molten metal into a pre-made mould. It is usually known as *die casting*, where the mould is in several pieces so it can be taken away from the hardened component made. Moulds are designed with feeder runs and overflow runners that have to be machined away from the finished casting. The design of these extra runs and the composition of the metal poured is critical to ensure a controlled size of grain during the casting process, and to prevent the formation of any blow-holes from furnace gases. Sand casting is the best known means of forming from molten metal and the reverse pattern is often made, ironically enough, from hand-crafted wooden shapes which are then pressed into sand, which may be made into a hard stable material by the addition of clays, which is then heated to 550 K. Alternatively, the silica sand is coated with sodium silicate which is then hardened by passing carbon dioxide gas through the mixture. The castings have to be finished mechanically as the final surface is rough.

Older methods use the technique of *investment casting* or *lost wax casting*, where the shape to be cast is made in wax which is coated with ceramic particles. The whole piece is then mounted in a box or tube and the remaining space is packed with coarser ceramic particles. The piece is heated and the wax runs out, leaving a hollow form. The whole piece is then fired to form a strong ceramic that can take the weight of the molten metal. Once fired the mould is broken away. It does not lend itself to repeat forms and is a labour-intensive process, but it does reproduce fine detail.

Centrifugal casting This spins molten metal into the inside of rotating moulds to produce pipes; no complex solid moulds are needed for this process as the metal solidifies as it cools while turning.

Pressure die casting This uses accurately made metal formers which are filled with molten metal usually under pressure. The mould is removed after casting and the process is designed for making thousands of similar objects, otherwise the whole process and the accurate machining of the dies would be uneconomic. Parts made from this process need a minimal amount of finishing work and complex shapes can be made including such fine detail as threads for bolts and integral lettering.

Extruding A solid piece of metal known as a *billet* is rammed through a die which reduces the cross-sectional area of the metal and forms it into a continuous section. Not only simple rods and pipes can be produced but also complex sections for windows and door frames. It is a hot working process. Very accurate sections can be produced. As a common analogy, it is like toothpaste issuing from a tube.

Forging Forging involves the pressing of metals into shape by use of considerable force. It is a hot working procedure and the shapes produced can be complex, determined by the shape of the die into which the metal is being pressed. The comparative flow of the metal is critical as too much stress will induce resulting stress patterns which are frozen into the final form. Stress concentrations will be apparent at corners. Crankshafts are made by this process, and pieces made in this way may need further stress relieving operations or annealing.

Sintering or powder fabrication Sintering involves the pressing of metals in powder form into a mould. Under pressure some cold welding between particles takes place through energy produced by friction, and the whole component is heated to allow grain growth to take place across the cold welds. Intermetallic materials use a mixture of metallic and ceramic particles to produce *cermets* which then have a wide range of applications including tungsten carbide tools, and other cemented carbides which have a high degree of abrasion resistance are useful for brake pads. Some specialist components need to be oil-less bearings. Mixtures of powdered copper and tin with graphite are sintered above the melting point of tin which, if quenched with lubricating oil, give a self-oiling bearing.

Joining by welding Welding involves the controlled heating and partial melting of metals so they fuse together. The technique has inherent problems in that the structure of the metal will be altered locally and can lead to weakness through recrystallization or local distortion. Gases used in welding may stay in solution with the metal leading to embrittlement. Small sections of metal may cool too rapidly, so leading to embrittlement. Welding needs good control and welders must be tested to current British Standards.

High energy sources are needed to raise the temperature sufficiently to melt the metal and a filler rod of a compatible alloy. This is supplied by electrical arc welding, flame from a gas torch as in oxy-acetylene welding, explosive welding or electron beam and laser welding.

Soldering and *brazing* are carried out at lower temperatures and require careful preparation of surfaces to maintain an intermetallic bond across the join without being weakened by thick oxide coatings. Brazed joints are stronger and can take higher stress than soldering, and are used in the making of bicycle joints for example. Soldering acts as a sealant and relies on the metals being held permanently in place so the joint will not suffer any stress.

Solid phase welding applies to the heating of two metals to be joined below the melting temperature and then applying mechanical pressure to make a join. This technique is used in spot welding and strip welding and also in riveting where the rivets are heated and then hammered, as in shipbuilding.

Mechanical joining, riveting, bolting One of the drawbacks of this method of joining metals is that it involves making an inherent weakness in the metals to be joined by forming holes. Great stress will then be put on the joining mechanism and shear stresses on bolts have to be calculated. The machining of holes, which are already stress concentrators, has to be carefully controlled to avoid defects that could act as crack initiators under load.

Machining The action of machining uses a metal harder than the metal to be worked which is profiled in the form of a fine-angled wedge which will progressively remove layers of metal. It is not as simple as cutting as it requires an amount of physical energy to drive the wedge-shaped section which is called the *tool*. This usually remains static, while the metal to be worked is rotated or turned. There is local deformation as the edge of the metal is sheared from

Figure 6.2 Fire tender, New York City. This truck uses a steel body malleable enough to be pressed into complex curvatures. Durable paint finishes, chromed bumpers against corrosion, chequer-plate panels where wear is greatest and flush doors that have to be weatherproof are used.

Figure 6.3 Services stacks, Lloyds of London, Richard Rogers Partnership (1986). External services are commonplace in oil refining plants or other processing industries, and the choice of metals and finishes has to be precise to avoid unacceptable deterioration over the life of the building. Costs may be recouped by not having to build a complete weatherproof enclosure which requires its own foundations. As industry makes economical decisions, there are more reasons for taking this strategy rather than architectural expression.

Figure 6.4 One of two celebration towers built in 1989 for the bicentennial celebration of the French Revolution in Paris by Jean Marie Hennin and Nichols Normier with engineer Peter Rice. They are built of steel with fabric wings, housing exhibition spaces and a café. They show a vitality and elegance reminiscent of first flight.

the parent surface. Surfaces are then finished and polished to remove some of the deformation that occurs and also defects that may be crack initiators at a later time. Hard metals are machined, as softer metals will produce greater deformation in shear and an unsatisfactory surface finish. However, the metal to be machined has to have sufficient ductility to undergo some small local deformation otherwise it will fracture. The local deformation also acts as a miniature strain hardening mechanism to the surface. Local coolants are used to reduce the local high temperatures, and these are often emulsions of oil in water.

Recycling

The collection of scrap is used in the recycling of iron–carbon alloys, and all other metals for secondary processing. Domestic scrap and automobile scrap, recycled for steel, contain high amounts of impurities, particularly copper, and only the heavier and pure engineering scrap is of value in charging the furnace for steelmaking. The quality of scrap deteriorates, however, as the same steel is in circulation, and there is an accumulation of impurities which affect the quality, and ultimately the grade of steel, limiting its future use. The most valuable recycling commodity for steel in this country is stainless steel scrap. Although there are presently good supplies of steel, iron ores are a diminishing resource, and heavy use will affect the price if there is a limited supply on the world market.

The more complex the alloyed structures used, the greater the difficulties in recycling (due to the gain of impurities that are too expensive or impossible to extract, which all result from earlier steelmaking). It is worthwhile trying to maximize and make efficient the use of steel. It is ironic that recycling is somewhat similar to the original processing of the material, except that the input of energy multiplies every time the scrap makes a circuit. Although there is this large energy input in reprocessing scrap, even the production of steel has different efficiency ratings according to the end product made. For example, steel pressings have a higher wastage than rolled steel sections from ingots, as waste material has to be collected and remelted, taking greater energy. Machined parts have a higher wastage than cast parts, again producing a high percentage of waste that needs to use energy in recycling. Designers should be conscious of the implications in specifying steel in terms of the energy value attached to production methods. Ultimately decision making will have to be influenced by the longevity of the building or component. Short-life structures should use steels and finishes that can be easily recycled, whereas permanent structures could use more complex alloys.

Figure 6.5 Imprimerie Mame, Tours, Jean Prouve with the architect Henri Prouve (1950). This design in aluminium cladding on a steel structure provided a complete roofing and glazing element that was repeated down the lengths of the shed. The deep double skin form had sufficient rigidity to span just over 4.0 m.

6.6 Designing with metals

The structural capability of metals is important even for the smallest component. Their usage is highly specific and, as the demand for careful engineering architecture increases, much can be learnt from other technologies, whether used in the automotive or aircraft industries for example. Transport engineering has already had to cater for the development of weatherproofing details that are efficient to keep occupants dry and water out of engines.

The work of Jean Prouve exemplifies an integrated approach to design that makes simultaneous decisions in materials, structure and form. The resultant shape combines Prouve's innate understanding of the material used in

Figure 6.6 Jean Prouve was always interested in the double cantilever roof beam structure. He used folded steel sheet to make up structural elements reflecting the zones of maximum stress by increasing areas of metal at critical junction.

Figure 6.7 Chairs and desk developed for school children using folded steel structures.

Figure 6.8 Stainless steel canopy, Savoy Hotel (1933). One of the prime concerns in specifying metals is durability in exposed situations. Initial outlay with high performance metal coatings or alloys last.

conjunction with a full understanding of the processing methods needed to achieve that form. This produces economies in the use of sections employed; as the engineering becomes more minimal, reflecting the function need of a particular element, it also becomes more elegant. Jean Prouve represents in one man what normally has to be a successful partnership between designer and manufacturer, where experience of handling and making artifacts is shaped by a dialogue reflecting the purpose of the finished piece. These illustrations are from the exhibition on Jean Prouve held at the Pompidou Centre in 1991.

6.7 Health

In *The control of substances hazardous to health in the construction industry*, fumes and gases used in welding are noted as a hazard. Welding, brazing and cutting are the chief activities that produce fumes and gases from metalworking, resulting from the different metals and fluxes used. These fumes affect the respiratory system by causing major irritation. The chief gases involved are carbon monoxide, nitrous oxides and ozone. There should be controlled extraction local to the operations, with good ventilation. Helmets should be used.

Lubricants for machinery based on mineral oils can cause dermatitis, acne and possibly skin cancer, and in aerosol form give respiratory damage. Hazards to health from particular metals are mentioned in the individual sections (see lead details, page 145).

References

1 *Statistical Yearbook* United Nations 1987.

7 Ferrous alloys

7.1 Iron

Iron was used as a structural material in the eighteenth century with the Iron Bridge at Coalbrookdale in 1779 as the first large-scale structure in cast and wrought iron. The early use of the material exploited the knowledge of timber technology and jointing which, to some extent, made sense as castings originated from timber formers. Cast iron columns which were either hollow round sections or cruciform flanges continued to be used until the early part of this century. The *Building in steel, A.J. Guide* 31st August 1983, contains some useful information on working with these kinds of existing structures, including evaluation of their strength from information of the period.

Wrought and cast iron

Wrought iron is a pure iron which is refined pig iron from which impurities, such as carbon, silicon and sulphur, have been extracted usually by oxidation through manipulation of the solid material. In making wrought iron there are always some slag particles remaining and the characteristic of wrought iron is that it is forge welded, hammered so that further slag particles are removed, but those that are left are formed into long fibres which can be seen clearly under the microscope running across ferrous grains. The material has then directional strength. Hammering at high enough temperatures is a form of welding as there is crystal growth across the layers of metal. One of the problems in early iron production was in achieving high enough temperatures for slag to run freely as a molten material, and temperatures must be between 1675 and 1775 °C for this to happen. When the refinement of pig iron started to use air blown through molten pig iron, slag could be directly removed without the laborious mechanical process of forging. The first low carbon steels were produced using the Bessemer *converter*, controlling the composition of the metal while it was in a liquid state. As a nearly pure iron it is very malleable and tough, so resistant to impact damage made wrought iron ideal for chains for heavy marine use. It is also more resistant to corrosion than steel due to the formation of an oxide coating which is fairly homogeneous.

Wrought iron does have reasonable tensile strength and was a major building material until the costs of making low carbon steel were lowered. Doubts were raised over its material quality in the catastrophic Tay Bridge disaster of 1979 when the bridge, a mixture of wrought and cast iron components, was subjected to great stresses. Although actual material failure was found in the cast iron elements, steel was then chosen in preference for its greater strength in subsequent civil engineering structures. One of the last major engineering projects was the Tower for the 1889 Paris exhibition by Gustave Eiffel which used 7300 tons of wrought iron. Wrought iron is still used for the complex shaping of decorative ironwork and because of its durability is ideal for external work.

The composition of slags in iron and steel making has to be controlled so that impurities can be removed efficiently and elements that need to be left in iron and steel are retained. Slag is a mixture of metal oxides and silica and has the important function of helping to control the oxygen supply to the metal and also acting as a thermal barrier, keeping the liquid metal at a constant temperature.

7.2 Steel

Steel is a widely used material and architects and designers use it in a variety of situations from structural elements and lightweight cladding components, to furniture, ironmongery and fixings. Each different application puts different demands on the material needed. This is achieved by alloying or processing giving a wide range of steels with very different properties. It is in the designers's interest to grasp some of the technology associated with steelmaking.

Figure 7.1 Small cast iron bridge made by the Boston Lodge foundry (1854). Cast iron relied on hand-crafted joinery for making the positive forms that were then impressed into sand moulds. Consequently, this bridge reflects the pattern-making characteristics of early Victorian timber detailing.

Figure 7.2 Gate by Gaudi at Parc Guell, Barcelona (1900–1914). In Catalonia, the forging of iron is part of the history and culture of the area. This gate makes reference to the shapes of beetle wing cases in the use of curved profiles.

An understanding of the origins of the steel, and the energy and processing techniques to refine the material are indicative of the way in which steel can corrode. Analysis of this process gives better strategies in learning how to cope with preventing decay, either by specifying more durable alloys or by choosing and specifying coatings and other protective techniques.

Manufacture of steel

Steel is the only alloy in general use which is known by a name that is descriptive of its qualities, rather than being derived from the elements of the alloy. Steel has an iron–carbon composition and its name derives from *stahl* meaning firm or rigid. It is a material with a degree of resilience, i.e. loading can be taken with some deflection without incurring brittle behaviour (a characteristic typical of ceramics for example). All metals are generally *tough*, a term used to describe ability to resist the propagation of cracks, which means they will deform plastically prior to fracture, and show some resistance to impact damage. This is in contrast to ceramics which will reach their elastic limit and then fail.

Steel is derived from iron and carbon. Iron is extracted from iron ores that are mined from the earth's crust, and part of the refining process is to remove oxygen from these ores by heating them with coke and limestone to a temperature of about 1600 °C in a blast-furnace.

In ironmaking there is a series of complex reactions between the charge of coke (carbon), limestone (calcium carbonate), and the various ores (iron oxides). Gas (carbon dioxide) is produced from the decomposition of limestone (carbon carbonate), and another gas (carbon monoxide) is produced from the chemical combination of coke and oxygen. Both of these gases take part in an interchange to reduce iron oxides to iron. There can also be a straightforward reaction between iron oxide and hydrogen to produce pure iron and water. All of these reactions are temperature dependent and require careful control by the

Figure 7.3 'The Wall' by Coates Branson, Tokyo (1990), manufactured in England (see Figure 5.4 for the whole building). The cast iron screen shows the advantage of casting forms to facilitate junctions.

blast-furnace managers to ensure that an economic reaction (in terms of fuel used) takes place.

The iron oxide is either reduced directly by the oxide combining with carbon to form iron and carbon monoxide, or indirectly by combining with this newly formed carbon monoxide to form iron and carbon dioxide. After this process there is always some carbon left in the solid iron formed, and this amounts to about 4% by weight. The material is known as *pig iron* in this state with an overall content of 6% of other unwanted elements, including silicon, manganese, sulphur and phosphorus in unmeasured quantities, even though these elements may be required in alloying later.

The pig iron must be further refined to reduce the carbon content and the other unwanted elements before the material can be categorized as a particular steel, as cast irons, with about 4% carbon, are brittle and cannot take large tensile stresses.

There are impurities in the melt in steelmaking which have to be removed before the correct specification can be achieved. These include phosphorus and silicon (which makes steel hard and induces brittleness), sulphur (which can cause cracking in poured castings and also over-hardening) and carbon which could take the material into the unwanted categories of a cast iron. This range of iron carbon steels are referred to as *plain carbon steels* as the only other element present is manganese, present in up to 1.6% weight maximum, and left in the metal from de-oxidation and de-sulphurization processes. The addition of any other elements will bring the metal into the metallurgical category of *alloy steels*. This has an immediate effect on the price of steel by affecting availability. Orders for steel ideally relate to the tonnage that can be processed in one batch and this can limit specific alloy applications to very large projects.

Refining requires the re-melting of the iron in a steel-making furnace with a large oxygen input. The refined molten steel is cast into ingots and these are reheated before subsequent reprocessing into slabs or blocks, depending on the final profile needed. Continuous casting methods are used where demand can justify this large-scale production. Molten steel is poured into the top of a water cooled mould and the steel is pulled through while still solidifying before further cooling and rolling into the final cross-section.

Steel as an alloy

There are more alloys of iron than any other metal used, and steel covers a large range of ferrous alloys, with some additional elements, depending on the final performance needed from the material, or the demands made on it in processing. The chief alloying elements are iron and carbon, and although the percentage amounts of carbon may seem small, because of the relative density differences between the elements, the volumetric contribution of carbon is large and must be calculated precisely in order to control the microstructure of the material.

Most steels have carbon contents far lower than 2.11%. This figure marks the critical maximum where in the processing of the material, at 900 °C, a complete phase change can occur, giving steel its characteristic properties of great strength. Above this level of carbon content, an iron–carbon alloy becomes more brittle taking on the type of performance associated with cast irons. The manipulation of the carbon content is carried out by the controlled removal of carbon by oxidation, introduced in the early stages of iron production.

Although phosphorus, silicon and carbon can be regarded as impurities if over a desired percentage in the melt, they are desirable as alloying elements in the right quantities.

Up to 0.15% weight phosphorus can harden steels to aid accurate machining. Nickel and chromium improve strength, and in low chromium steels, 3% molybdenum is often added to improve impact values. Nickel, chromium, silicon, copper and aluminium also improve corrosion resistance. If carbides can form (e.g. chromium, tungsten, vanadium, molybdenum, niobium and titanium carbides), as their strength is greater than iron carbide they become important in engineering for cutting tools. All of these different elements affect the *microstructure* of the metal. In building there is also a demand for heat resisting steels used for studs and bolts, etc., so that in their attachment to other steel elements by welding, their inherent microstructure and predicted strength are not affected.

The general term *alloy steels* in the industry refers to those steels which have a 5% or more by weight contribution from other elements. Table 7.1 gives some typical steel constituent quantities.

The fundamental property which distinguishes steel from other metals and its parent metal iron, is its great strength. Steel can also be modified to cater for a particular usage. By manipulating the carbon content, steel can be engineered to have a wide range of properties and its strength can be improved whereby it can show a high strength to weight ratio for construction, and can be shaped to accommodate different geometries and structures. Depending on the section design it can be used in compression or tension and the structural strength can be easily predicted. Structural connections can be easily achieved whether by welding or by bolting.

Structural and non-structural steels

These two major categories of structural and non-structural steels are defined by carbon content.

Structural steels that carry designed loads are described as *medium carbon steels* with typically 0.12–0.24% carbon. They describe all the basic steels used for engineering, often rolled steels. Non-structural steels fall above and below this category.

Steels below 0.15% are described as *low carbon steels* and are flexible enough for pressing into sheets for cladding and being drawn into wire. Because of the flexibility of these steels (which do enable shapes to be pressed into three-dimensional shapes) they are relatively weak, and for panels and sheets to have some minor structural value, the shapes pressed into them are also designed to improve their rigidity. These shapes are also designed to be stress-relieving, i.e. they have no right angle junctions. To improve stiffness they may become part of a composite panel system that achieves rigidity by depth of construction. Adhesion technology can then become an important part of the overall system. Steels between 0.5 and 1.5% carbon are described as *high carbon steels* and are used for casting as well as machining. Casting steels for special applications, i.e. joint details for nodal connections between structural elements, need good quality control and during their manufacture should be checked (often by X-ray or ultrasonic equipment) in case there are inclusions which have to be made good by welding.

The appropiate specification of steel is dependent on usage. The grade of steel numerically relates to the minimum tensile strength expected for that grade. For example, a grade 43 steel will have a minimum tensile stress of 430 N/mm^2 and a grade 50 steel a minimum tensile stress of 500 N/mm^2. There is some adjustment to grades

Table 7.1 Typical steel constituents

Steel type	Constituents	Applications
Low carbon mild steel	0.04–0.3 C + 0.8 Mn	Low stress constructional steel suitable for welding
Medium carbon steel	0.3–0.7 C + 0.8 Mn	Medium stress uses, machinery, etc.
High carbon steel	0.7–1.7 C + 0.8 Mn	High stress uses, springs, cutting tools
Low alloy steel	0.2 C + 0.8 Mn + 1.0 Cr + 2.0 Ni	High stress uses, pressure vessels, aircraft parts
High alloy steel (stainless steel)	0.1 C + 0.5 Mn + 18.0 Cr + 8.0 Ni	High temperature or anti-corrosion use
Cast iron	1.8–4 C + 0.9 Mn + 2.0 Si	Low stress uses, cylinder blocks, rain pipes

Source: M. Ashby and D. Jones 1986 *Engineering materials 2* Pergamon Press: p. 4

Table 7.2 Steel grades and designations

British Standard		Minimum tensile strength (N/mm^2)	Minimum yield strength at 12 mm (N/mm^2)
4360:1990	7668:1993		
Grade	Designation according to EN 10027-1 and IC10 (see note)		
WR 50A	S345JOWPII	480	345
WR 50B	S345JOWII	480	345
WR 50C	S345GWII	480	345

Note: The steel name is derived from EN 10027-1 and IC 10 as follows:

- the letter S for structural steels
- the indication of the minimum specified yield strength for thickness ≤ 12 mm expressed in N/mm^2
- the characters JO to indicate an impact requirement of 27 J at 0 °C
  or
- the letter G to indicate an impact requirement of 27 J at − 15 °C
- the letter W indicating that the steel has an improved atmosphere corrosion resistance (weather resistant)
- the letter P for the class with a greater phosphorus content
- the letter H to indicate that the product is a hollow section

Source: Table A.1 BS 7668:1994

in the 1986 version of BS 4360 to bring tensile strength groups into bands more compatible with European standards. The general rule of grades relating to tensile stress holds true except for grade 40 which now has a minimum strength of 340 N/mm^2. New steel designations to be adopted are given in Appendix H of that standard and the nearest equivalent grades to those in Europe are given in Appendix A (reproduced in Table 7.2). The full range of weldable steels are described in BS 4360 *Specification for weldable structural steels*. There are four major tensile strength ranges: grades 40, 43, 50 and 55. Every strength grade has a further categorization between A and E reflecting a change in alloy specification, and a higher structural grade that is tougher and less likely to undergo brittle failure (important in welding at low temperatures). Advice will be needed for specialist applications particularly where high strength is required, i.e. for cables and their connections and where there is a high level of exposure. It is common for architects and engineers to work directly with the steel manufacturers in these situations.

There are differences in steel prices depending on the quality needed and the different alloys specified. As a guide, high yield steels are about 10% more expensive than mild steels, and stainless steel can be seven times the price of ordinary steel. (Information from British Steel)

7.3 Processing and microstructure of steels

Although the properties of steel can be controlled by the alloying elements used, they can also be altered by the heat treatments carried out, the method of cooling used and the method of processing. All these factors affect the crystalline structure of steel (referred to as the *microstructure*). By carefully controlling the microstructure, the properties will be predetermined and this will ultimately affect the specific use of the material.

Hot rolled sections

Hot rolling is, in effect, hot working the material and some defects, such as voids in the original cast metal, are removed in the rolling process. Most of the structural sections used by architects and engineers are hot rolled. White hot ingots of cast steel are sent into the rolling mill and are passed through sets of rollers which gradually change the profile into our recognizable I and H sections. These are known as *universal beams and columns* due to a single process which deforms the metal in two directions. These sections are identified by serial size and mass per unit length. Each serial size corresponds to a different set of roller sizes. Changes in mass per unit length are achieved by increasing the thickness of the flanges during rolling.

BS 5950 1990 *Structural use of steelwork in building* is the current standard relating to hot rolled sections.

Cold rolled sections

Although some structural sections are cold rolled, it is more likely that this process is used for smaller structural sections used in lattice trusses, etc. Small sections used in cladding rails may be cold rolled or drawn, and cladding panels with profiles are produced by rolling steel sheet into shape. This will produce sections that have undergone some crystal deformation and will be harder and stronger than hot rolled sections. Cold rolled processes are often referred to as *work hardening* processes, and more malleable alloys are sometimes used to prevent excessive stress on the metal if complex shapes are being made, for example, stainless steel sinks. Alternatively, slow cooled steel is more ductile, and will be used for complex pressed shapes, such as car bodies. The shape of the section being made will also be designed to have slightly curved profiles and rounded internal corners, which are stress relieving shapes if the metal is being deformed under pressure. Cold drawing is the process used to manufacture tubes, rods and all wire, and as the section sizes reduce, there is interim annealing to the metal, relieving stress between different stages of drawing.

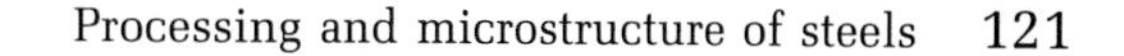

Figure 7.4 Structure supporting the airship hanger doors at Cardington, Bedford (1909). Typical of cold-rolled sections. The doubling up of slender sections gives stiff yet lightweight elements within the three-dimensional configuration of the bracing structure.

Rolling tolerances

Although modern rolling mills can achieve a relatively good degree of accuracy and consistency, it is essential to have certain rolling tolerances. It is more critical for hot rolled materials. This is because of roll wear during rolling, as well as some degree of temperature variation and mechanical wear on mill components. Although these tolerances generally have very little effect when the products are used in normal fabrication and in structural applications, they should be considered when detailing structures which have components needing a close fit, and yet are manufactured with more precise dimensional co-ordination. Allowance must then be made for connections which have slots for adjustment, and the tolerances given to a cladding manufacturer, for example, must take into account this kind of inaccuracy on the steel structure. Co-ordination with the steel fabricator at the design stage will give the architect the correct information. (The dimensions and tolerances for structural steel sections are covered by BS 4 Part 1:1972.)

Steel castings

Steel casting has more applications in heavy engineering than in building but they may provide the best solution to making a complex shape in three dimensions. There may be a need to design special joint connections which can be used to provide difficult intermediate knuckle joints in structures resolving the connection of one element to another. It is also possible to cast in stainless steel with chromium contents varying from 13 to 21% for corrosion resistance.

Sheets and strips

These are important elements in building and are covered by BS 1449 Part 1 1983: *Steel plate, sheet and strip* (refer also to Part 2). Nominal widths are 600 mm but can be wider and up to 16 mm in thickness with nominal lengths of 3000 mm. Tables 7.3 and 7.4 give symbols for surface finishes and surface inspection, and material condition; BS 1449 also contains grades and alloy composition information. Sheets can be rolled into deep sections that provide roofing sheeting and also the decking for permanent formwork receiving a structural concrete topping in fast track building construction.

Cooling and heat treatments

The most dramatic transformations can be achieved in steel by fast cooling or *quenching*, which can be achieved in oil. Instead of the normal crystal structure, small needle-shaped crystals of martensite form, creating a great deal of stress inside the metal. This makes the metal extremely hard but also brittle.

To demonstrate this effect, if a flexible rubber sheet is placed in liquid nitrogen this will have the effect of supercooling the material into a rigid glass like sheet. It will then shatter into fragments if struck. The brittle fracture from embrittlement by rapid cooling shows very quickly that processing and temperature control does have an effect on the internal structure of materials.

Metal components are subjected to this treatment if they need to be hard for a particular purpose, such as for tools or fixings. Some tools such as chisels or masonry nails have different heat treatments, and are quenched at their tip where hardness is needed, but can deform under stress at

Table 7.3 Symbols for surface finish and inspection

Finish	Symbol	Description
Pickled	P	A hot-rolled surface* from which the oxide has been removed by chemical means
Mechanically descaled	D	A hot-rolled surface* from which the oxide has been removed by mechanical means
Full finish	FF	A cold-rolled† skin passed material having one surface free from blemishes liable to impair the appearance of a high class paint finish‡
General purpose finish	GP	A cold-rolled† material free from gross defects, but of a lower standard than FF‡
Matt finish	M	A surface finish obtained when material is cold-rolled† on specially prepared rolls as a last operation
Bright finish	BR	A surface finish obtained when material is cold-rolled† on rolls having a moderately high finish. It is suitable for most requirements, but is not recommended for decorative electroplating
Plating finish	PL	A surface finish obtained when material is cold-rolled† on specially prepared rolls to give one surface which is superior to a BR finish and is particularly suitable for decorative electroplating
Mirror finish	MF	A surface finish having a high lustre and reflectivity. Usually available only in narrow widths in cold rolled material
Unpolished finish	UP	A blue/black oxide finish; applicable to hardened and tempered strip
Polished finish	PF	A bright finish having the appearance of a surface obtained by fine grinding or abrasive brushing; applicable to hardened and tempered strip
Polished and coloured blue	PB	A polished finish oxidized to a controlled blue colour by further heat treatment; applicable to hardened and tempered strip
Polished and coloured yellow	PY	A polished finish oxidized to a controlled yellow colour by further heat treatment; applicable to hardened and tempered strip
Vitreous enamel	VE	A surface finish for vitreous enamelling of material of specially selected chemical composition
Special finish	SF	Other finishes by agreement between the manufacturer and the purchaser

Note: Some of these finishes are available only from wide mills, others only from narrow mills.
* A hot-rolled surface supplied in the as-rolled condition, i.e. a non-descaled surface; some surface defects of steelmaking or rolling origin may be present, the proportion being greater with strip than with sheet or plate.
† By special agreement between the supplier and the purchaser, cold-rolled material can be supplied to surface texture specifications. The information to be given in statements about surface texture requirements is specified in BS 1134: Part 1.
‡ FF and GP finishes are normally supplied with a matt appearance.

Source: Table 2 BS 1449 Part 1:1983

their head where they are hit, otherwise they can fracture in a brittle manner causing injury. It is normal practice to *temper* some steels to reduce some of the stresses, and to achieve hardness as well as *springiness* to avoid brittle failure during use. The martensite transforms to a different crystal structure as the steel is tempered by heating to between 300 and 600 °C. This critical temperature is not high enough to start complete recrystallization, and so the metal retains a great deal of its original structure and strength.

In contrast, slow cooling will produce a metal that has very large crystals (which have time to grow) and which are said to be equiaxed, i.e. they have the same linear value in all three dimensions. *Stress-relief annealing* is carried out sometimes after the metal has had its crystal structure distorted and overhardened by either quenching or cold work processing.

Stress-relief annealing allows the inbuilt stresses to be relaxed to allow further work to take place. This is achieved by heating the metal to a point where there can be some re-crystallization (usually at about 0.6 of the melting temperature of the metal), and for steel this will be about 650 °C, and then cooling. If the cooling is not assisted and the steel is cooled in air, this is known as *normalizing*. At this stage the metal has greater ductility and can easily be bent into the required shape. After a shape has been formed, the metal can be re-heated and quenched, putting the metal into a hard but brittle state.

Re-heating until 250 °C is reached and then cooling in air will temper the metal, making it tough and springy or *tempered*.

Tubular furniture or bicycle frames are a good example where metals have to be heat treated to give properties of strength without embrittlement, otherwise they would not be able to take normal everyday stresses. In a tubular chair a fully annealed version would be unable to take load without excessive deformation, and a fast cooled version would be brittle with a higher elastic limit, uncomfortable

Table 7.4 Symbols for material conditions

Condition	Symbol	Description
Rimmed steel	R	Low carbon steel in which deoxidation has been controlled to produce an ingot having a rim or skin almost free from carbon and impurities, within which is a core where the impurities are concentrated
Balanced steel	B	A steel in which processing has been controlled to produce an ingot with a structure between that of a rimmed and a killed steel. It is sometimes referred to as semi-killed steel
Killed steel	K	Steel that has been fully deoxidized
Hot-rolled on wide mills	HR	Material produced by hot rolling. This will have an oxide scale coating, unless an alternative finish is specified (see table 2)
Hot-rolled on narrow mills	HS	
Cold-rolled on wide mills	CR	Material produced by cold rolling to the final thickness
Cold-rolled on narrow mills	CS	
Normalized	N	Material that has been normalized as a separate operation
Annealed	A	Material in the annealed last condition (i.e. which has not been subjected to a final light cold rolling)
Skin passed	SP	Material that has been subjected to a final light cold rolling
Temper rolled		Material rolled to the specified temper and qualified as follows:
	H1	Eighth hard
	H2	Quarter hard
	H3	Half hard
	H4	Three-quarters hard
	H5	Hard
	H6	Extra hard
Hardened and tempered	HT	Material that has been continuously hardened and tempered in order to give the specified mechanical properties

Source: Table 1 BS 1449 Part 1:1983

to use and liable to fracture. The first experimental tubular chairs made by Breuer in 1925 were developed from bicycle sections but:

'The first version of the armchair was a source of consternation for Breuer. The welding of the pre-bent pieces made the chair very stiff; it had no resilience, a quality he considered essential for comfort.' (*Marcel Breuer furniture and interiors*, Christopher Wilk, The Museum of Modern Art, 1981.)

The final version of the chair by Thonet in 1929 had a continuous tubular steel structure, and the idea was developed in the fully cantilevered B32 Thonet side chair (1928). This was very similar to the cantilevered chair by Mart Stam and a legal dispute led to a court observation on the nature of the metal as the Stam chair was made of 'laquered, cast (non-resilient) steel tubing' and that the Breuer-Thonet model B33 was made from 'nickel plated precision steel tubing'. (*Marcel Breuer furniture and interiors*, p. 76.)

Mart Stam was eventually given credit for the original design. As the shape was so similar, the difference in material and properties was thought inconsequential, yet the flexible behaviour of the Breuer chair gave real comfort.

Although these are small-scale examples, they do show that manipulation of the material can produce changes in yield strength and breaking strength. On a small scale it is easier to appreciate differences between metal treatments and behaviour. On a larger scale it is more difficult to see material behaviour, for example, deflection over large spans, and we have to rely on calculation for prediction.

Welding

Welding is a neat and strong way of jointing structures as the method of joining ensures that two pieces of metal are effectively fused and act as one structurally. Welding is a highly skilled activity and the method has to be carefully specified to British Standard 5135. Provision should also be made for the testing of welds in highly stressed elements. Welding is also used to repair defects in steel except for round or square bars.

As steel undergoes such controlled transformations during

Figure 7.5 Element of Miami House by Coop Himmelbau erected as part of an exhibition by the Architectural Association. Steel fabrication in Germany: (left) with flat plate it is possible to weld up quite complex geometries, and provide fixing points with local stiffening (right).

processing, it is essential that the microstructure of the material is not destroyed by poor jointing methods. Welding, which uses heat to melt steel locally, can alter adjacent microstructures. Embrittlement through over-ageing can occur if the weld cools too slowly. This is why welding must be carried out to approved standards, to ensure that there is no loss of strength across the weld, and it is vital to specify welding standards in building documentation. Sometimes unwanted intermediate alloys can form in the weld and the higher the carbon or other alloy content, the more difficult it will be to weld. If there is over 0.54% carbon in the steel, martensitic structures can form in the weld zone from fast cooling, and these can make the weld brittle. It can also encourage hydrogen cracking which is a repercussion of hydrogen gas from atmospheric moisture forming small voids in the weld. Some high strength steels may have to be pre-heated prior to welding, which can help avoid embrittlement from fast cooling.

As welded areas cool, stresses can be generated which interfere with bonding, ultimately causing cracking.

Welding can be carried out by:

- Gas, which uses a hot flame from a torch to melt the parent metal in conjunction with a welding or filler rod.
- Electric arc, which uses a current in conjunction with a filler rod
- Fusion from electron beam welding.
- Explosion from detonated charges.
- Spot, which uses electrodes either side of metal sheets to be joined.

All these methods use high amounts of energy, resulting in heat to locally melt and fuse two pieces of metal, at the same time ensuring continuity of the metallic bonding and crystal structure across the join. As welding effectively makes a larger piece of the same material, there are repercussions in having one large piece. Defects can spread across large areas of material and some components have to be designed so that large sheets are joined by riveting. Cracks cannot then jump across separate components.

Weld testing Critical structural welds should always be tested and there are various methods of doing this. A visual inspection can be made to see that there is sufficient material across the weld (to allow for grinding down later if necessary), and that there are no obvious defects, pin-holes, etc. Visual inspections can be enhanced by using special dyes, sometimes fluorescent, to show defects more clearly. The metal is cleaned, dried and then coated with a film with a penetrating fluid containing the fluorescent compound. The surface is flushed clean and any faults show up clearly in the metal where the fluorescent compound has collected. Welding defects can also be shown up by magnetizing the metal component to be tested. After spraying the welds to be tested with a white paint, a dark magnetic powder is oversprayed and the powder collects either side of the crack, acting as an indicator.

The ultrasonic testing of welds is a very accurate way of pinpointing defects. Pulses will normally be reflected from the bottom edge of a piece of metal where there is a change in medium. If there is a defect, the pulses will not cross the defect and will be reflected back earlier. Automated testing machines running along welds can indicate major faults which are immediately coated with paint markers to record where further work is needed. X-ray detection methods are also used to identify faults.

The Liberty boats constructed from 1941 as supply ships in the Second World War ultimately showed a failure in terms of their design. They suffered from brittle fracture which was a consequence of the type of steel used (which had a low manganese:carbon ratio) and a detailing problem which put a great deal of bending stress in a localized situation near a hatch. Their welded structure meant that a crack could propagate throughout the whole structure of the ship. Out of 2700 ships some 400 were affected, with 90 suffering major damage, and several broke completely in half. The brittleness of the steel would also have increased with the very low temperatures in the North Atlantic.

Under some conditions steel may exhibit a brittle rather than a ductile mode of failure. This is not normally a problem in building structures since modern steels are manufactured in such a way as to avoid such behaviour. However, in extremely cold conditions, for instance exposed steelwork where temperatures start to fall below 20 °C, this is known as the *transition temperature* and the brittleness of steel will increase rapidly. Steels under stress with high carbon contents are more vulnerable. In these low temperature conditions steel alloys are carefully specified to cater for these situations, often the manganese: carbon ratio is increased and the nickel content reduced. Specialist advice must always be sought. Steel in these conditions becomes notch sensitive, so related fixing details that can impose defects on steel from screw threads, etc. must be avoided.

Figure 7.6 Transfer structure, Broadgate, by Ove Arup & Partners (1991). This junction shows the use of heavy steel plate which is welded and riveted and mounted on a cast base. The making of junctions has to use a variety of techniques to achieve an elegant solution.

7.4 Steel corrosion

Steel is derived from iron, which in turn is extracted from iron ore. Iron is derived from some fairly complex oxides, sulphides and carbonates. For example, the taconite ores are haematite ($2[Fe_2O_3]$) and magnetite ($8[Fe_3O_4]$). The extraction of metals and the energy needed to carry out the processing is linked to the ultimate deterioration of all metals.

As steel is not found naturally and is artificially engineered, it is in a state recognized as being thermodynamically *unstable*. It will readily combine with oxygen and moisture to exist in a state of lower free energy. In striving to achieve a more stable configuration chemically, it reverts to a state reminiscent of its original form as an iron oxide prior to mining and mineral processing. This is the basis of corrosion.

Once iron and steel are made, care has to be taken to protect them in the environment, to ensure that there is a minimization of the thermodynamic tendency to change their state in the presence of moisture and oxygen and, in our terms, *rust*.

Although there is a variety of systems/techniques to protect steels from corrosion, their specification will be dependent on their situation and degree of exposure.

The prevention of corrosion by alloy specification

See *MBS*: *Finishes* for more details.

Copper and its alloys oxidize in the atmosphere and form

Figure 7.7 La Geode at Parc Villette, Architect A. Fainsilber, Engineer Peter Rice (1986). The stainless steel cladding has open joints with a weatherproof skin behind. An unexpected benefit is the self-cleaning effect of the polished surface, nothing can adhere to it. The point fixing of flat sheets of metal is problematic; without continuous support they will ripple.

a stable patina, which if scratched is self-renewing, and this gives them inbuilt corrosion resistance. Structural steel can be alloyed so that it can form similar protective surface layers which eliminate the need and cost of applied coatings.

Chromium in steel forms a strong passive chromium oxide film that is protective, and when containing more than 12% chromium steel is given the common name of *stainless steel*. Stainless steels should contain at least 12% chromium, but this is an expensive solution to corrosion problems in the building industry where the cost of large quantities of this metal is often prohibitive. Consequently, the use of stainless steels is restricted to highly exposed components, i.e. brick support angles and external fixings, or those that receive so much mechanical wear, i.e. handrails or ironmongery, that this higher level of specification can be justified in the long term. Specialist advice should be sought for installations in particularly corrosive environments, such as chemical works, swimming pools, or marine environments.

Weathering steels may be seen as an alternative in providing a self-finished material suitable for external use. Weathering steels (categorized as WR 50 A1 — Cor Ten A and WR 50 B1 — Cor Ten B) have a higher percentage of copper (up to 0.50% by weight) and phosphorus (up to 0.15% by weight) than normal steels. Typically they contain 1.5–2.5% alloying elements principally copper, chromium, nickel and phosphorus, which facilitates the formation of a more stable oxide layer. The finish of these steels is typically brown–purple but there are problems attached. Different batches of steel might have varying alloy contents, but differences in weathering between very exposed surfaces and sheltered areas, can produce a noticeable difference in the colour of the oxide layer. Rainwater will also be discoloured after washing over these steels and can stain adjacent paving and concrete fascia details. External finishes should then be carefully chosen, particularly at the base of the building. Some successful details use a wide margin of similar coloured aggregates around the perimeter of a building for drainage. This strategy is also referred to as *controlled patination*. Full development of a protective coating can take over two years, and clients should be warned of the metamorphosis of their building.

Both chromium steels and weathering steels can corrode at a rate equivalent to mild steel in marine or equally aggressive environments and so specialist advice should be sought prior to making final decisions on the alloys to be used.

Defects can arise from using large areas of sheeting and one characteristic failure is star-cracking at the centres of sheets flexing from negative wind pressures. It is cyclic loading that eventually leads to fatigue. Problems occur in using sheeting over 3 m in length. There are also condensation problems and ventilation is still needed. It is important to avoid contamination from other metals, such as steel fixings and cast iron downpipes. (Source: BRE Digest 121, *Fully supported stainless steel roof coverings*, RICS publication, 1985. Michael Watford.)

Instead of the classic polished appearance a grey finish of low reflectivity can be used. The kind of sheeting is normally 28 gauge (0.375 mm) laid in bays up to 9 m. The jointing of sheets can use standing seams or wood rolls; details must allow for thermal movement.

Figure 7.8 Lloyds of London, Richard Rogers Partnership (1986): the cladding on the service towers is linen-finished stainless steel. The quality of concrete is of an extremely high standard, the formwork used is plastic-faced plywood. Twenty different mixes for the concrete elements were tested in the process.

Figure 7.9 Tokyo Gymnasium, Fumihiko Maki (1990). This design owes its development to the Fujisawa Municipal Gymnasium 1984 which had a similar roof constructed of 0.4 mm stainless steel sheet. 'The process of design and construction were one' taking two years to work out and two years to assemble. It has references to Samurai helmets and giant molluscs, and the stainless steel roof reflects the sky, at times almost disappearing, one of Maki's intentions.

Figure 7.10 Temple roof, Ueno, Tokyo (under restoration, 1990). The folding of copper sheet into complex shapes for roofing is one of the traditions of Japan and a precedent for the Tokyo gymnasium.

Figure 7.11 An unusual use of galvanized steel as large format panels with pressed decorative relief on an ordinary house in Le Thilot, edge of Jura, France. The coating has decayed and the panels are now rusting, but after a long life.

Coatings for corrosion control

MBS: *Finishes* includes greater details of corrosion control coatings.

If metals have no other protective mechanism they must be coated for protection. Coatings can be achieved by using paint systems, metal coatings (as in galvanizing), or fused powder coatings that may be inorganic and form glass ceramic coatings. Whatever system is finally chosen, surface preparation is the most important part of the coating process, and poor cleaning and preparation is usually the cause of most paint adhesion failures.

Surface preparation Steel supplied from the mill has surface mill scale which should be removed first by blasting, but this surface cleaning treatment should be applied to all steel prior to coatings being applied. The general sequence is to blast the surface of the steel, prime and finish. Blast cleaning can be achieved by grit blasting or shot blasting. Shot blasting is preferred as it gives an even series of indentations with a rounded profile and does not suffer from possible contamination problems from impurities present in grit blasting. The depth of blasting must relate to the finish to be achieved, being generally one third the depth of the full coating system. Ideally the surface should be cleaned after blasting, the surface dried at 500 °C for 90 minutes, and then priming carried out immediately, preferably within two hours.

The right conditions for cleaning and coating surfaces are in the steelwork fabricator's premises. Any site-applied finished will not be as satisfactory and should be avoided where possible. Manual methods on site will use wire brushes and scrapers and should still be specified by referring to the Swedish Standard SIS 055900.

Figure 7.12 Corrugated iron which is galvanized has a long life, especially in dry climates here at Sofala in New South Wales, and is easily replaced.

Paint coatings Paint coatings usually comprise binders that carry pigments and they may also contain solvents. They harden either by oxidative polymerization (which is by combination with the oxygen in the atmosphere to form what is known as *convertible coatings*); or by the evaporation of solvents (known as *non-convertible* coatings). Coatings should be seen as an effective barrier to water and air. They must have good properties of adhesion to the substrate, be consistent through their thickness and be proven in terms of their performance. Their depth is measured in microns, and a single coating thickness may only be 25 microns (the thickness of a sheet of paper.) As reliance is so heavy on the integrity of coatings, they must be carefully specified. The range of coatings on the market is vast and a result of polymer engineering for very specific applications. This is why manufacturer's recommendations must always be followed and a traditional coating system, consisting of primers, undercoats and top coats applied in the correct sequence. There are other coating systems which are applied as a sprayed powder, which adhere electrostatically to metal substrates and are then fused onto the surface of metals in ovens. These epoxy or polyester resins often carry guarantees but are conditional on regular maintenance which includes cleaning cycles of between three and six months depending on the location.

Any coatings used will only be fully effective if the preparation is adequate. The specification of the preparation of metal is important to ensure there will be adequate adhesion or bonding for the coating. In specifying the coating system, this will depend on the environment as well as the quality and depth of the coating finish. The actual coating thickness will increase from approximately 70 microns to over 450 microns in the most vulnerable environments.

Reference should be made to BS 5493:1977 *Protective coating of iron and steel structures against corrosion*, BS 7079:1990 *Specification of blast cleaned steel for painting*, and CP 3012:1972 Code of Practice for *Cleaning and preparation of metal surfaces*. There are the steelwork Corrosion Protection guides published by British Steel in association with the BCSA (British Constructional Steelwork Association), the Paintmakers Association and the Zinc Development Association. These main guides cover *Building interiors*, *Exterior environments* and *Cavity walls* (which is useful for all enclosed situations). The information given specifies the degree of cleaning and preparation, the type of coating which ranges from organic coatings and micaceous iron oxides to sprayed metal finishes (zinc and aluminium), full galvanizing and ethyl zinc silicate preparations.

Metal coatings Steel can be protected from water by coating with another metal. Galvanizing, which is the dipping of steelwork into a bath of molten zinc, is the cheapest and most common method. A zinc–iron alloy is formed at the steel surface which gives perfect adhesion and total impermeability to moisture. Galvanizing is particularly cost effective for light structural steelwork, lattice girders being a good example.

Metal spraying, using a specially designed spray gun to throw molten metal particles onto a clean and roughened steel surface, is used for both zinc and aluminium coatings. The metal droplets freeze into the surface and form a slightly porous layer of overlapping plates which should be impregnated with a sealing treatment to give maximum corrosion resistance. No alloying takes place and adhesion is achieved by mechanical bonding to the specially roughened surface. The cost of sprayed coatings is high because of the extreme demands of surface preparation and cleanliness. If this coating is damaged both zinc and aluminium will continue to protect the underlying steel from corrosion by galvanic action, and both metals will give very good performance in clean neutral conditions.

Zinc will also perform well also in mildly alkaline conditions although subject to chemical attack in acid environments. Aluminium is unsuitable for use in alkaline conditions. Both metals can be overpainted, though special primers are required, and this, together with suitable maintenance schedules gives an extremely long life to steelwork.

All these coating systems have different lifespans. Paint films can become embrittled through ultraviolet light in the atmosphere and in their life have to cope with great extensibility of the substrate and need a programme of renewal. Sometimes it is more problematic to renew coatings that initially had a long life and were originally one-coat heat treated and factory applied applications. There are problems in specifying *in situ* treatments that have good adhesion to these original coatings.

Protection by detailing and good site practice

Poor detailing can accelerate corrosion. All exposed surfaces of steelwork should have weathered details which do not encourage the retention of water and allow for drainage. (See BS 5493 Code for *Protective coating of iron and steel structures against corrosion*.) Simple principles should be followed and channel sections, for example, should always be used facing down rather than up. If the detailing unavoidably collects some water, there must be provision for drainage. Any horizontal ledges will collect particles which can assist in starting corrosion. Bolt ends, nuts, etc. must be detailed for protection and not allow water to collect, i.e. domed heads should be used externally. Joints should be overlapped ensuring water will flow over and not into joints. Steel on site must be protected preferably under cover, stacked above ground level and well ventilated.

All of these protective measures are preventative, as corrosion is an inevitable change of state, whether by direct oxidation or electrochemical corrosion, which we try and delay, or at least inhibit to give a reasonable working life for the material. However, well detailed and protected steels have a long working life and this can be extended by regular inspection and maintenance.

8 Non-ferrous alloys

8.1 Aluminium

Aluminium is a metal which has seen the greatest growth and usage in the last 50 years since the Second World War. From its application in high-level engineering in the aircraft industry a number of products have been designed which use its high strength:weight ratio. It is the most abundant metallic element as a resource in the earth's crust, and at 7.5% is the second most abundant solid element after silicon. In its natural form it is found in strong chemical combination with oxygen and a great deal of electrical energy is needed to release it. Consequently its processing is best sited adjacent to renewable energy sources to generate electricity, and its situation next to hydroelectric power stations is ideal. The strength of bonding in aluminium oxide means it is insoluble in water, and for this reason it is an important and strong constituent in cement and in refractory materials for lining furnaces. Hard crystalline forms of aluminium oxide are found as sapphire (with blue impurities) and ruby (with red impurities). The controlled engineering of aluminium oxides results in very strong materials which can be used as cutting blades. When the oxide is refined it produces a pure soft metal which is not used structurally. Various alloyed combinations are able to give an increase in strength 30 (83,000 psi) times greater than in its pure state (6,500 psi). As a pure metal it is used for foil and coatings in the building industry.

One great landmark in aluminium and engineering was the development of the Burlington Zephyr train built in 1933. For the first time an aluminium diesel engine had been developed that was light enough to mount on a moving locomotive frame (previously they had to be permanently mounted on bases). The alloy used was *duralumin*, developed in Germany and used first in coinage and then for the frame of early zeppelins. The alloy contained 3.5% copper, and 0.5% magnesium with existing traces of iron and silicon. The alloy was quenched in water after annealing but after four or five days strengths were shown to have increased due to age hardening. At room temperature aluminium cannot hold more than 0.5% of the copper atoms in solid solution, and they migrate to areas of pure aluminium to form the intermetallic compound $CuAl_2$.

Table 8.1 Standard properties of aluminium

Property	Value
Atomic weight	63.546
Atomic number	13
Density	2700 kg/m^3
Coefficient of linear expansion	24×10^{-6}/K
Thermal conductivity	226 W/m K
Melting Point	659.7 °C
Tensile strength *pure*	90 kN/m^2
Youngs Modulus	70 GN/m^2

The cladding for flying boats showed another early use of alloys in aluminium that needed great strength, achieved by a very stiff alloy containing 45% copper, 1.25% magnesium, and 1 % nickel. After age hardening, its strength improved by 50%. This alloy is susceptible to corrosion in marine atmospheres and, for this reason, was protected by being rolled with pure aluminium on either side to form a sandwich construction.

Aluminium is low in the electrochemical series and considered one of the base metals which implies it would corrode at a great rate. However, it oxidizes very quickly in the atmosphere and this oxidized layer, although only 10^{-6} mm in thickness (1 micron), is sufficiently well bonded to form a totally non-porous layer which is impervious to water. This natural process of oxidation is enhanced by anodizing which gives a controlled thickness of 5 microns of this oxidized layer for internal finishes, and up to 25 microns for external use. This oxidized layer can be engineered to give different colorations by the effective thickness. This controls the wavelengths of light and subsequent colour rendering which can be further controlled by dyeing the oxidized layer to give a range of colour finishes.

Classification of alloys

In processing aluminium it is divided into two major groups: wrought alloys (BS 1470–1475) and cast alloys (BS 1490) which have different alloy compositions. Each of these major groups is further subdivided into heat treatable alloys or non-heat treatable alloys. Generally all heat treatable alloys are strengthened by using ageing mechanisms and all non-heat treatable alloys are strengthened by solid solution strengthening, strain hardening or dispersion strengthening.[1] BS 1470 was amended in 1989 to give different designations and harmonize with ISO 2107.

In the wrought and cast forms of aluminium alloy materials prefixes are:

N Non-heat treatable alloy
H Heat treatable alloy

The second letter denotes the material form:

S Plate sheet and strip BS 1470
T Drawn tube BS 1471
F Forging stock and forgings BS 1472
R Rivet bolt and screw stock BS 1472
E Bars extruded round tubes and sections BS 1474
G Wire BS 1475
B Bolt and screw stock BS 1473
R Rivet stock BS1473

Additional suffix letters give the material condition after processing:

F Material as fabricated, no formal heat treatment
O Fully annealed to give lowest strength
T Generally refers to thermal treatment to produce stable tempers
H Refers to being strain hardened

H1 to H8 give grades of cold-working after annealing, the higher the number the greater the strength and hardness. As eight is the maximum hardness that can be achieved, two will indicate $\frac{1}{4}$ of the hardness that can be achieved.

T Refers generally to being thermally treated
T3 Solution heat treated, cold worked and then naturally aged
T4 Refers to solution treated and then aged naturally
T5 Cooled from an elevated temperature which has shaped the metal and then subjecting it to precipitation hardening (artificially aged)
T6 Solution treated and precipitation hardened
T8 Solution treated, cold worked and precipitation hardened

If these materials are heated then their strength will be reduced unless they are subjected to further strain hardening.

Table 8.2 gives the foreign equivalents of UK aluminium alloys. As the use of components is now international this information is useful to give recognition to the equivalent alloys used.

Wrought alloys which are not heat treated have good strength and rigidity with good corrosion resistance. They are used in automobile and marine engineering and architecture, providing durable panels and superstructures in marine environments. There is a great increase in the use of aluminium for offshore structures, such as oil rigs. One of the latest structures designed for the Norwegian oil industry is for a five-storey accommodation block weighing 2000 tonnes with extrusions that are up to 100 mm thick. Ancillary structures to this project include helidecks for helicopter landings and link bridges back to the main oil platform.

Wrought alloys which have been heat treated exhibit age hardening or precipitation hardening. Here, slow cooling allows the intermetallic compound $CuAl_2$ to move within the atomic lattice causing local distortion, restricting movement and increasing the stiffness of the material. They are used for components which have more structural purposes including glazing bars for the automobile and building industry, stressed parts of aircraft structures, ladders, scaffolding tubes, and tubular furniture. It is the strongest commercial alloy produced in the industry.

Figure 8.1 Institut du Monde Arabe, Paris, Jean Nouvelle (1987). The aluminium grid holds 30,000 diaphragms which adjust to the change in light quality externally. The alloys used are the same specification for high strength components in the aircraft industry.

Table 8.2 Alloy cross-references

BS alloy designation	Old BS alloy designation*	ISO alloy designation	BS temper designation	Old BS temper designation
Unalloyed				
1050A	E1B	Al 99.5	F	M
1200	E1C	Al 99.0	F	M
Non-heat-treatable				
5083	NE8	AlMg4.5 Mn	O F	O M
5154A	NE5	AlMg3.5	O F	O M
5251	NE4	AlMg2	F	M
Heat-treatable				
2014A	HE15	AlCu4SiMg	T4 T6	TB TF
6060	Not previously in BS 1474	AlMgSi	T4 T5 T6	TB TE TF
6061	HE20	AlMg1SiCu	T4 T6	TB TF
6063	HE9	AlMg0.7Si	O F T4 T5 T6	O M TB TE TF
6063A	HE9	AlMg0.7SiA	T4 T5 T6	TB TE TF
6082	HE30	AlSi1MgMn	O F T4 T5 T6	O M TB TE TF
6463	Previously BTR E6 in BS 4300/4	AlMg0.7SiB	T4 T6	TB TF

* The 'old' BS alloy designations were those used prior to the issue in June 1980 of amendment No. 1, which introduced the 4-digit alloy designation system.

Source: Table 18 BS 1474:1987

Cast alloys which are not heat treated are used in general sand and die castings. Complex sections can be produced and they have good corrosion resistance. Rigidity is more important than high tensile strength and cast alloys generally contain high proportions of silicon (between 9 and 13%). Cast alloys which are heat treated have high strength if subjected to high working temperatures, such as in engines for cylinder heads and pistons.

Generally, cast and wrought alloys use copper, magnesium, silicon, manganese and zinc. Casting alloys will use tin in addition but not manganese.

Aluminium alloys for structural purposes This standard (BS 1161:1977) covers sections that relate to equal and unequal angles, channels, I sections T sections, equal and unequal bulb angles, channels, lipped channels and bulb tees. Table 8.3 gives details of aluminium alloys suitable for structural purposes. (Refer also to the 1983 amendment to BS 1161.) Table 8.4 shows the chemical composition of heat-treatable aluminium alloys.

General alloy groups in aluminium have categories which are four-numbered, with the first digit representing the major alloying element.

Table 8.3 Aluminium alloys suitable for structural use

Material designation	Condition	0.2% proof stress minimum (N/mm^2)	Tensile strength minimum (N/mm^2)	Modulus of elasticity ($\times 10^3$ N/mm^2)	Minimum elongation (%)		Mass conversion factor compared with density of 2.71 kg/dm^3 of pure aluminium
					on a gauge length of $5.65\sqrt{S_o}$	on a gauge length of 50 mm	
5083	O	125	275	68.9	14	13	0.982
	M	130	280		12	11	
6063	TE	110	150	65.5	8	7	0.996
	TF	160	185		8	7	
7020	TB	190	300	71.7	12	10	1.026
	TF	280	340		10	8	
6082	TF	255	295	68.9	8	7	0.996
5454	O	85	215	68.9	18	16	0.989
	M	100	215		16	14	

Source: Table 1 BS 1161:1977

- Aluminium 99% minimum and greater 1000
- Aluminium alloying groups:
  - Copper 2000
  - Manganese 3000
  - Silicon 4000
  - Magnesium 5000
  - Magnesium and Silicon 6000
  - Zinc 7000
  - Other 8000
  - Unused 9000

The 1000 group is composed of pure aluminium used for sheet rolling and foil making. Pure aluminium has a high resistance to the passage of gases and moisture, which is why it is the obvious choice for the packaging of pharmaceutical products or as a barrier for moisture penetration in building. At over 25 microns in thickness it is regarded as impermeable. It also reflects 90–95% of radiant heat and so provides an effective thermal barrier in buildings.

The strongest group of alloys is the 2000 series which are copper-based and the most common engineering alloy is 2014A used in aircraft engineering. The high copper content gives poor weathering characteristics and it is still clad with a thin layer of pure aluminium on either side. This alloy is also used for drinks cans. The chemical composition of alloy 2014A, as an example of the alloying elements used, is given in Table 8.5.

Table 8.2 gives Alloy and temper designation cross referenced as the 1987 standard assigns numbers instead of letters to alloys and their method of treatment. It also gives full explanation of the coded letters used for alloys and their system of treatment.

The 3000 series is the manganese group, and these alloys have a small addition of manganese at 1%, and provide increased strength yet retaining the durability of aluminium. The products made are malleable enough for roll forming and ideal for profiled sheet cladding and panelling generally. A new use is for satellite receiver dishes. These alloys are particularly suitable for the formation of tight bends.

The 5000 series has magnesium as the main alloying element and these alloys are used for rolled sheet and plate products with high strength from cold working and good corrosion resistance in marine conditions. They are used for panels in naval engineering. In the transport industry, their relatively low weight (a third of steel) and their high strength makes them obvious choices for the fabrication of heavy duty transport vehicles, such as dumper trucks, with all-aluminium chassis and bodies. For increased stiffness they will have characteristically deep strengthening sections. They also have important applications in the production of polymers where they are used for die making and vacuum injection. Proof stress is 125 MPa.

The most common alloys used, especially for ordinary engineering applications and extrusions in the UK are in the aluminium–magnesium–silicon group or the 6000 series. The main alloying categories are 6063 (H9) and 6082 (H30). The advantage of this group of alloys is that they can be processed into complex shapes while ductile, before natural age hardening gives them increased stiffness. They can also be used for hardened plate products which may be stamped with a three-dimensional chequer plate patterning which gives extra support for walkways, or floors for milk delivery floats. Proof stress is 115 MPa.

The 7000 series which use zinc as the main alloying element produces higher strength alloys used in military

Table 8.4 Principal properties of aluminium alloys

Alloy	Condition	Product	Thickness (mm)		Minimum 0.2% tensile proof stress (N/mm^2) (see note 1)	Minimum tensile strength (N/mm^2) (see note 1)	Elongation (%) (see note 1)		Approximate loss of strength due to welding (%) (see note 2)	Durability rating	Related British Standard (see note 3)	Old BS alloy designation	ISO designation (nearest equivalent) (see ISO 209-1)
			Over	Up to and including			$5.65\sqrt{S_0}$	50 mm					
6061	T6	Extrusions*	—	150	240	280	8	7	50	B	BS 1474	H20	AlMg1SiCu
	T6	Drawn	—	6.0	240	295	7†	7†	50		BS 1471		
		Tube	6.0	10	255	295	9†	9†	50				
6063	T4	Extrusions	—	150	70	130	16	14	0	B	BS 1474	H9	AlMgO,7Si
	T4	Drawn tube	—	10	100	155	15†	15†	0		BS 1471		
		Forgings	—	150	85	140	16	—	0		BS 1472		
	T5	Extrusions	—	25	110	150	8	7	25		BS 1474		
	T6	Extrusions	—	150	160	185	8	7	50		BS 1474		
	T6	Drawn tube	—	10	180	200	8†	8†	50		BS 1471		
		Forgings	—	150	160	185	10	—	50		BS 1472		
6082	T4	Extrusions	—	150	120	190	16	14	0	B	BS 1474	H30	AlSi1MgMn
		Sheet	0.2	3.0	120	200	—	15	0		BS 1470		
		Plate	3.0	25	115	200	12	15	0				
	T4	Drawn tube	—	6.0	115	215	12†	12†	0		BS 1471		
			6.0	10			14†	14†					
		Forgings	—	150	120	185	16	—	0		BS 1472		
	T6	Extrusions	—	20	255	295	8	7	50		BS 1474		
			20	150	270	310	8	—	50				
		Sheet	0.2	3.0	255	295	—	8	50		BS 1470		
		Plate	3.0	25	240	295	8	8	50				
		Drawn tube	—	6.0	255	310	7†	7†	50		BS 1471		
			6.0	10	240	310	9†	9†	50				
		Forgings	—	120	225	295	8		50		BS 1472		
7020	T4	Extrusions	—	25	190	300	10	12	0	C	BS 4300/15	H17	AlZn4,5Mg1
		Sheet and plate		25	170	280	10	12	0		BS 4300/14		
	T6	Extrusions	—	25	280	340	8	10	40		BS 4300/15		
		Sheet and plate		25	270	320	8	10	40		BS 4300/14		
LM25	TB7	Sand cast	—	—	80 to 110	160‡	2.5		NA	B	BS 1490	LM25	AlSi7Mg
		Chill cast	—	—	90 to 110	230‡	5						
	TF	Sand cast	—	—	200 to 250	230‡	—						
		Chill cast	—	—	220 to 260	280‡	2						

Notes:
1. Minimum value specified in BS 1470, BS 1471, BS 1472, BS 1474 and BS 4300/14 and BS 4300/15.
2. Properties after 3 days natural ageing at 15 °C to 25 °C for 6* * * series alloys and 30 days at 15 °C to 25 °C, or artificially aged for 24 h at 120 °C, for alloy 7020 (see 4.4.2.1).
3. For chemical compositions of alloy see relevant British Standard.

* Extrusions refers to bars, extruded round tubes and sections.
† Elongation on $5.65\sqrt{S_0}$ for tubes with wall thickness of 3 mm and thinner, and on 50 mm for tubes with thicker walls.
‡ Minimum value specified in BS 1490.

Source: Table 2.1 BS 8118 Part 1:1991

applications. This type of alloy is also used in moulding plates for the vacuum injection plastics and for moulds generally. Proof stress is 170–200 MPa.

The specification of aluminium relies on the correct selection of two parameters: the appropriate alloy and the right method of heat treatment or cold working to optimize the performance of that alloy for a particular use. Information from manufacturers of individual products often leaves out the actual alloy designation but will use the classification for treatment to indicate the mechanical properties. For the detailed specification of aluminium and its alloys refer to *The properties of aluminium and its alloys* published by the Aluminium Federation, 1983.

Table 8.5 Chemical composition of aliminium 2014A alloy

Element	Composition (%)
Si	0.5–0.9
Fe	0.5
Cu	3.9–5.0
Mn	0.4–1.2
Mg	0.2–0.8
Others	1.0
Aluminium	Remainder

BS CP 118 *Structural Use of Aluminium* 1969 is still used as an indicator by some manufacturers for the overall performance of their products.

Principal structural alloys

- N8 welded structures and platework also can be extruded into simple structural sections. These are very durable in marine environments.
- H9 extruded solid or hollow sections have good durability and surface finish and are good for anodizing. They are used for complex extrusions.
- H30–TF is used for welded structures and platework and can be extruded to simple structural sections. H20 is similar.
- H15 has special applications usually in the aeronautical industry. It is less resistant to crack propagation.
- N3 is used as NS3–H8 is used for corrugated and troughed sheet in wall and roof cladding systems. It has reasonable durability.

Tables 8.6 and 8.7 give properties of the principal alloys and their permissible stresses.

Temperature control is a factor and modification is required where structures will exceed working temperatures of 66 °C (i.e. in most buildings).

At joints of aluminium to zinc or galvanized steel and to steel, cast iron and lead there is a recommendation to follow the protection instructions given in CP 143 *Sheet and wall coverings* Part 15 Aluminium:1984. This code recommends that underlays are used as isolators between aluminium and other sheet materials. In some situations bitumastic paint can be used as an isolator between aluminium and steel or zinc-rich coatings.

Production of aluminium

Aluminium can be rolled as plate with a thickness over 6 mm and up to 200 mm, or as sheet which is below 6 mm but above 0.2 mm. Strip is coiled sheet with thicknesses between 0.2 mm and 3 mm. Foil is rolled at thicknesses below 0.2 mm although it is more likely to be as thin as 0.0006 mm. In the process of rolling, patterns can be embossed to give extra stiffness, as well as a patterning to provide a more durable surface.

Many sections are formed through the extrusion process where the ductility of aluminium is used to advantage and the metal can be rammed through a die into solid or hollow sections. These sections are often profiled to give a complex section which stiffens the material where it is subject to the greatest stress. This allows aluminium to be saved and metal to be used efficiently.

Shapemakers As the technology for producing extruded sections has improved, there is a subsidiary service of the Aluminium Federation which represents the 14 main extruders in the UK, enabling special profiles to be made at very little extra cost. Computerized techniques allow for the production of sample extrusions within the short timescale of four weeks. A special designers pack gives

Table 8.6 Limiting stress for weld metals

Filler		Limiting stress (N/mm^2)								
Type	Alloy	Parent metal								
		Non-heat-treatable						Heat-treatable		
		1200	3103 3105	5251	5454	5154A	5083	6063	6061 6082	7020
1	1080A	55	—	—	—	—	—	—	—	—
	1050A	55	—	—	—	—	—	—	—	—
3	3103	—	80	—	—	—	—	—	—	—
4	4043A	70	90	—	—	—	—	150	190	—
	4047A	70	—	—	—	—	—	150	190	—
5	5356 5056A	—	—	200	215	215	245	155	205	255
	5183 5556A	—	—	200	215	215	275	165	220	265
	5554*	—	—	—	190	—	—	—	—	—
	5154A*	—	—	—	—	210	—	—	—	—

* These alloys should be used in conditions where corrosion is likely to be a problem.

Source: Table D.1 BS 8118 Part 1:1991

Table 8.7 Limiting stress for heat-treatable aluminium alloys

Alloy	Condition	Product	Thickness (mm)		Limiting stress (N/mm²)		
			Over	Up to and including	p_o	p_a	p_v
6061	T6	Extrusion	—	150	240	260	145
	T6	Drawn tube	—	6	240	265	145
			6	10	225	260	135
6063	T4	Extrusion	—	150	65	85	40
	T4	Drawn tube	—	10	95	120	60
	T4	Forgings	—	150	80	100	50
	T5	Extrusion	—	25	110	130	65
	T6	Extrusion	—	150	160	175	95
	T6	Drawn tube	—	10	180	190	110
	T6	Forgings	—	150	160	170	95
6082	T4	Extrusion	—	150	115	145	70
	T4	Sheet	0.2	3	115	145	70
	T4	Plate	3	25	105	140	65
	T4	Drawn tube	—	10	105	140	65
	T4	Forgings	—	150	115	145	70
	T6	Extrusion	—	20	255	275	155
			20	150	270	290	160
	T6	Sheet	0.2	3	255	275	155
	T6	Plate	3	25	240	265	145
	T6	Drawn tube	—	6	255	280	155
			6	10	240	275	145
	T6	Forgings	—	120	255	275	155
7020	T4	Extrusion	—	25	185	230	110
	T4	Sheet, plate	0.2	25	160	205	95
	T6	Extrusion	—	25	280	310	170
	T6	Sheet, plate	0.2	25	270	295	160

Source: Table 4.1 BS 8118 Part 1:1991

a strategy for the design of sections to be used in the drawing office. This can be obtained from Shapemakers Information Service, Broadway House, Calthorpe Road, Five Ways, Birmingham B15 1TN.

Foreign alloy classifications Table 8.8 shows the nearest foreign equivalent to designated wrought and cast alloys.

Aluminium and fire

Although aluminium melts at 660 °C (a lower temperature than steel), and its thermal conductivity is high, the surface reflectance is also high, so it can reflect heat which can be taken into account when calculating the overall thermal resistance. Fire wall structures can be successfuly designed to give two-hour fire protection. The structural capability of aluminium is seriously reduced when temperatures exceed 480 °C. However, it does not burn in air or give off flammable vapour or gases, and satisfies conditions for being non-combustible.

Welding and joining in aluminium

Welding should use tungsten arc welding (TIG) or metal arc welding, including the pulsed arc (MIG) inert gas process. The protective aluminium oxide coating has to be melted at the initially high temperature of 2000 °C before the aluminium alloy can be successfuly welded. No flux is needed but a shielding arc of argon gas must be used to prevent the re-oxidation of aluminium in the formation of the weld which would otherwise cause embrittlement. The correct welding wire should be used according to the alloy of the aluminium to be connected.

Technology has advanced to use laser welding techniques (used mostly in the microengineering needed for the electronics industry but spreading into general engineering applications). Explosive welding and friction welding, known as *solid phase bonding techniques* are used for joining to dissimilar metals, such as copper alloys or steels. This category of joining metals also includes ultrasonic techniques and cold pressure processing for the joining of foil and other thin sections.

Table 8.8 Nearest foreign equivalents to BS specified alloys

British Standard	ISO	Former BS	International AA*	Country				
				Austria	Belgium	Canada	France	Germany
1200	Al99,0	SIC	1200	Al99	1200	2S	1200	A199
3103	AlMn1	N3	3103	AlMn	—	—	—	AlMn1
3105	AlMn0,5Mg0,5	N31	3105	—	—	—	3105	AlMn0,5Mg0,5
5083	AlMg4,5Mn0,7	N8	5083	AlMg4,5Mn	5083	5083	5083	AlMg4,5Mn
5154A	AlMg3,5(A)	N5	5154A	—	5154	—	5154	—
5251	AlMg2	N4	5251	—	5251	—	5251	AlMg2Mn0,3
5454	AlMg3Mn	N51	5454	—	5454	5454	5454	AlMg2,7Mn
6061	AlMg1SiCu	H20	6061	—	6061	6061	6061	AlMg1SiCu
6063	AlMg0,7Si	H9	6063	—	6063	—	—	—
6082	AlSiMgMn	H30	6082	—	6082	—	6082	AlMgSi1
7020	AlZn4,5Mg1	H17	7020	AlZn4,5Mg1	7020	—	7020	—
LM5	AlMg5Si1	LM5	514.1	G-AlMg5	SGAlMg6	GS40	A-G6	G-AlMg5
LM6	AlSi12	LM6	A413.2	G-AlSi	SG AlSi12 DG AlSi12Fe	S12N	A-S13	G-AlSi12
LM25	AlSi7Mg	LM25	A356.2 356.2 357.1 A357.2	—	SG AlSi7Mg	SG 70N SG 70	A-S7G A-S7G03 A-S7G06	G-AlSi7Mg

* Aluminium Association

Source: Table A.1 BS 8118 Part 1:1991

Traditional techniques of mechanical riveting can still be employed and these should use aluminium rivets or steel rivets if they are protected by an isolating sealing paint. Brazing is still used in some situations but forms a weaker joint which is more susceptible to corrosion.

Adhesion Although adhesion bonding is an important part of aluminium and composite fabrication technology, it is not covered in CP 143 Part 15 except by reference to an engineer. Jointing should not be done in the field, but be carried out under controlled conditions in the fabrication workshop. It requires the meticulous preparation of the joining surfaces and is used chiefly in the aerospace industry.

Components

As aluminium is important in the use of particular components, it is more relevant to look at guiding standards for those components which would include: cladding, windows and doors.

BS 4873 1986 deals with aluminium windows and it is worthwhile drawing attention to the type of alloy which could be from:

6063TF, 6063TE, 60663TB complying with BS 1474

Sills and other components, which do not require the same degree of structural stiffness as window or door frames, are fabricated from a different alloy. This requires ductility in pressing but not the same structural stiffness and alloys quoted are:

1200, 3103, or 5251 to comply to BS 1470 or alloy 5005 complying with BS 4300/7

Other alloys can be used providing they comply with the standards required for those that fall below a minimum thickness of 1.2 mm.

The method of finishing aluminium is specified to be one of three options.

(1) Anodising to BS 3987
(2) Liquid organic coating to BS 4842
(3) Powder coating to BS 6496

Corrosion standards and advice on electrolytic action between dissimilar metals are given. Fixings which will be in contact with the aluminium should be:

- Stainless steel grade A2 A4 or F1 to BS 6105.
- Steel that has been zinc plated to preferably Zn10 or Zn3 of BS 1706, chromate passivation to 2C or 2D of BS 6338.
- Hot dip galvanized to BS 2569 Part 1.
- A mechanical strength alloy of aluminium equal to the frame specification anodized to AA15 of BS1615 meeting 4.3 of BS 4873.

Italy	Japan	Norway	Spain	Sweden	Switzerland	USA
P-Al99,0	A1200	—	L-3001/Al99	Al99,0	Al99,0	1200
P-AlMn1,2	—	AlMn1	L-3811/Al-1Mn	Al-Mn1	AlMn	3103
—	A3105	—	L-3831/Al-0,5MnMg	—	—	3105
P-AlMg4,5	A5083	AlMg4,5Mn	L-3321/Al-4,5Mg	Al-Mg4,5MN	—	5083
—	A5154	—	L-3392/Al-3,5Mg	Al-Mg3,5	—	5154
P-AlMg2Mn	—	AlMg2	L-3361/Al-2Mg	—	—	—
P-AlMg2,7Mn	A5454	—	L-3391/Al-3MgMn	—	AlMg2,7Mn	5454
P-AlMg1SiCu	A6061	—	L-3420/Al-1MgSiCu	—	—	6061
P-AlSiO,5Mg	A6063	AlMgSi	L-3341/A1-0,7MgSi	Al-Mg0,5Si	AlMgSi0,5	6063
P-AlSi1MgMn	—	AlSi1Mg	L-3453/Al-1SiMgMn	Al-SiMgMn	AlMgSi1Mn	—
P-AlZn4,5Mg	A7M01	—	L-3741/Al-4Zn1Mg	Al-Zn4,5Mg	AlZn4,5Mg	—
G-AlMg5	ADC6 C7AV C7AS AC7A	AlMg5Si1	L-2331	4163	—	514.1
G-AlSi13	DIV DIS ADC1 C3AV C3AS AC3A	AlSi12(Fe)	L-2520	4261	G-AlSi13	A413.2
G-AlSi7MgMn	C4CV C4CS AC4C AC4CH C4CHS	AlSi7MgFe	L-2651 L-2652 L-2653	4244 4445	G-AlSi7Mg	A356.2 356.2 357.1 A357.2

The polymers for weatherstripping and gaskets are also specified.

- Weatherstripping
  - Chloroprene rubber solid (non-cellular to BS4255 Part 1).
  - Chloroprene rubber solid (cellular).
  - Cured ethylene propylene diene monomer (EPDM).
  - Polypropylene pile.
  - Plasticized PVC.
- Glazing materials (gaskets)
  - Chloroprene rubber solid (non-cellular to BS4255 Part 1).
  - Chloroprene rubber solid (cellular).
  - Cured ethylene propylene diene monomer (EPDM).
  - Uncured ethylene propylene diene monomer (EPDM).
  - PVC.

Advice is also given for glazing compounds and performance requirements in respect of weather-tightness in section 4.8 of BS 4873. One problem with metal frames generally is their thermal performance, and in order to produce an effective wall envelope, thermal breaks should be included if possible in the frame design. A range of polymers is used which are likely to be polyurethane resin, neoprene, uPVC, nylon, polyamide or rigid PVC foam.

BS 4873 is also relevant for doors and the Aluminium Windows Association document *Guidance in the handling, care protection, fixing and maintenance*, 1987, should be read in conjunction with this standard.

Attention should be drawn to maintenance and cleaning. Most manufactures carry guarantees for performance, but only if cleaning procedures are followed. Non-alkaline detergents and warm water are recommended and six-monthly cleaning is a minimum requirement which may be increased to three-monthly periods for polluted environments, such as industrial or coastal applications. Aggressive cleaners damage the surface of the aluminium irretrievably.

Cladding Roofing is usually recommended to be 1:40 although no less than 1:60. Batten roll and long strip systems can be used with a thickness of 0.8 mm and all details for the construction are given in CP 143.

Honeycomb sandwich panels These panels were developed for use by the aircraft industry and now have a wide range of applications which include panelling for other forms of transport, sports goods and building. The honeycomb core of the panel is made from a 3003 grade (manganese alloying group) aluminium foil only 0.04, 0.06 or 0.08 mm thick with a cell size of 5 or 6 mm diameter and weighing between 30 and 80 kg/m^3. For cladding, the honeycomb grade would be 32 kg/m^3 13 mm thick and

Figure 8.2 Kitagawara Arts Centre, Tokyo (1986). This building is called *Rise* as a reference to the aluminium sheet loosely folded over the end of the building to imitate curtain material. In reference to one of his latest buildings, Kitagawara says that he doesn't design his buildings to fit in with the neighbours as 'you never know what the neighbours will look like' in a place changing as quickly as Tokyo.

adhesive bonded to skins of 3003 alloy 1–1.5 mm. The honeycomb structure gives stiffness to the panel, preventing buckling or distortion. Surfaces can be polished textured, anodized, coated or laminated.

Finishes on aluminium

Smooth or bright finishes can be produced, ranging from light stain or matt or full matt. Although rougher surface texture surfaces can be produced, they are not generally encouraged as they reduce durability by causing dirt to be held within the surface roughness.

Aluminium will form a coating of aluminium oxide on exposure to air (a form of atmospheric corrosion) and this will gradually thicken to form a good protective coating. In a moist atmosphere the film produced will be white or grey, which can be rubbed off leaving a protective film underneath. Sulphates in industrial atmospheres and chlorides from marine atmospheres increase the rate of the film formed, which may be uneven, producing a series of small pits. In order to produce reliable protective film, finishing methods will either use anodizing as a controlled form of corrosion, or use applied films.

Weathering of aluminium and surface oxidation

- Light grey in rural areas
- Dark grey in industrial areas
- White matt oxide in coastal areas

Surfaces can be coloured by impregnating the anodic oxide coatings with organic dyes or pigments, the electrodeposition of pigments into coatings, and by choosing an alloy which may impart a different colour. The minimum thickness of an anodized coating should be 25 microns.

There is a system of *hard anodizing* which is used for saucepans giving them a characteristic grey–black finish which is extremely hard and abrasion resistant with the equivalent performance of chrome plating or stainless steel. Hard anodizing is carried out at low temperatures with dilute electrolyte solutions using high current densities and voltages. This gives a less porous anodized finish with greater hardness. The film can still be impregnated by pigments for colour or PTFE to reduce the coefficient of friction and give abrasion resistant surfaces. (See BS 3987:1974 (1982 amendment) *Anodic oxide coatings on wrought aluminium for external architectural applications*.)

Maintenance of anodized aluminium depends on cleaning and it is generally agreed that cleaning should be at a minimum of six monthly intervals and probably at shorter intervals depending on the degree of exposure. Manufacturers will often only give guaranteed periods of life if cleaning programmes are maintained and can be proved. Cleaning should only use water with a wetting agent or mild soap solution and only fibre brushes should be used. Any abrasive cleaning can damage the oxide coating. Wax polishes can be applied after cleaning. (See also *MBS: Finishes*.)

In BS 4842:1984 *Liquid organic coatings for application to aluminium alloy* pretreatment is emphasized with either alkaline or acid solutions. A chemical conversion coating is then applied first with either chromate or phosphate ions as the active components.

Protection to aluminium If aluminium is set into concrete, it should have a durability rating of A or B. When aluminium components are embedded in concrete there is a very slight reaction while the concrete is setting, which can etch and roughen the surface of aluminium. Components, such as lamp standards or flagpoles, which are to be embedded in this way should be protected by at least two coats of bitumen.

Protection is also needed where aluminium is set in contact with timber in industrial damp or marine environments. Care should be taken to specify timber that does not have a preservative treatment that uses copper sulphate, zinc chloride or mercuric salts. Oak, chestnut and western red cedar are again identified as particularly harmful because of their acid content. Water which runs off from copper sited locally will cause corrosion, so the close juxtaposition of materials needs to be considered when detailing junctions.

It should be noted that slender structural sections are more prone to torsional failure, and this helps explain the complex extruded sections in many aluminium sections to increase section stiffness.

Corrosion with other metals

Copper and its alloys, steel, nickel, lead and tin are cathodic to aluminium and so aluminium will suffer bimetallic corrosion if placed in juxtaposition with these metals, especially in industrial atmospheres and moist environments. If aluminium becomes saturated with water, any attack will be accelerated. Aluminium is less vulnerable to corrosion if in contact with lead or tin but, if situated in marine or industrial environments can still be affected.

Stainless steel and aluminium should form a stable combination in moderate atmospheres but corrosion to the aluminium in marine or industrial atmospheres can occur if both metals are totally immersed in water. The passive oxide coating to stainless steel does reduce the bimetallic effect.

Aluminium if alloyed with copper or zinc is more prone to corrosion. Zinc and magnesium or their alloys are anodic to aluminium and will corrode preferentially to aluminium, leaving the aluminium unaffected.

Table 8.9 Standard properties of copper

Property		Value
Atomic weight		63.5
Atomic number		29
Density		8.9 g/cm^3
Coefficient of linear expansion		16.8×10^{-6}/K
Thermal conductivity	*copper*	399 W/m K
	copper alloys	21–397 W/m K
Melting Point		1083 °C
Tensile strength	*copper*	215–25 MN/m^2
	bronze 60/40	325 MN/m^2
	phosphor bronze	455 MN/m^2
Young's modulus	*copper*	117–132 GN/m^2
	bronze 60/40	94–101 GN/m^2
	phosphor bronze	105–121 GN/m^2

8.2 Copper

Copper and bronze are metals that have been worked for at least 4000 years with evidence of pure copper being worked as early as 8000 years ago. Use of copper has changed from domestic implements and pigments, to artefacts of war, printing, and since the nineteenth century it has been fundamental to the electrical engineering industry because of its high conductivity.

There are over 160 copper minerals and copper is also found in a pure elemental state. It is more common to extract copper from sulphides which are found adjacent to mountain regions.

Applications

Approximately one third of all copper is used in the electrical industry. In building copper is used for roofing, water services (including supplies and heating systems), and also alloyed with tin to supply bronze fixings. Copper is also used in process engineering for distilling and food production. The most common alloys are: Brass, Bronze, Copper–nickel, Copper–silicon and Copper–aluminium.

Copper roofing See *MBS*: *External components* for details of roof coverings. Copper can be used for roof pitches of six degrees or greater. There is a basic requirement for falls to be 1:60. (See *Copper in roofing: Design and installation* published by the Copper Development Association, December 1985. This is a comprehensive book giving full details for traditional roofing which will use thicknesses between 0.34 and 0.7 mm and longstrip copper roofing which uses 0.6 mm thickness.)

Roofing components should use fully annealed copper to BS 2870:1980 section C104 or C106. Mechanical lapping and folding are traditional jointing methods which allow for expansion of the material. Rolls used as flat-seamed junctions can cause moisture penetration by capillary action.

Fixing materials such as nails should be copper or brass, screws should be of brass.

This is one of the main applications in the building industry used because it has a durable finish which naturally weathers and produces a stable corroded layer known as *patination* which is copper sulphate in urban areas and copper carbonate in rural areas. This should not be confused with *verdigris* which is the reaction of copper with acetic acid. Corrosion rates are given as microns/per annum (see Table 8.10) and patination can continue for 70 years.

The initial bright red surface develops an oxidized brown layer at first which then deepens to a darker brown before the secondary green patina develops. This patina development depends on the amount of sulphurous gases and chlorides present in the atmosphere and also on the inclination of the panels. Horizontal copper will change

Figure 8.3 The copper roof, its cuppola and fish weather vane sculpture has weathered to a rich turquoise on this rusticated stone control tower at the Craig Goch water reservoir in Wales.

faster, developing different shades of the green patina, but vertical surfaces will stay a dark brown. Artificial patination can be chemically induced but if the material is to be completely stable within the atmosphere it is better to let it weather naturally. It is common practice in the USA to rub raw linseed oil onto the metal surface which changes the deep red colour to a dark brown that will take about ten years to change to the green patina. Lacquering can be used to preserve the original copper colour and the International Copper Research Association has developed a lacquer coating called Incralac. Alternatively another coating known as Incracoat has been used which is a 1 mm thick sheet of polyvinylfluoride.

Underlays should be used that do not adhere either to the metal or any substructure, that allow the copper to slide under thermal expansion and isolate the copper from metal fixings used in the substructure. They should conform to

Table 8.10 Patination rates for copper

Atmospheric conditions	Patination	
	Rate (microns per annum)	Period (years)
City	1.5–2.9	8–12
Industrial	3.2–4.0	8–12
Marine	3.5	4–6
Country	1.9	up to 30

BS 747 Type 4B (ii) Brown sheathing (No 2 inodorous). (See CP 134 Part 2 1970 for substructures to sheet copper.)

Copper is also used for lightning protection. (See British Standard Code of Practice CP 326.)

Copper is stable if combined with lead or stainless steel and anodized aluminium. Pure aluminium will corrode preferentially to copper but the effects are minor and cause visual not structural defects. Mechanical isolation is still recommended to the differential movement between metals. Zinc and galvanized steel should not be installed below copper sheets in the direction of rain flow, and isolation should be with plastic foil or chlorinated rubbers. Bitumen will form corrosive acids due to environmental attack which can affect copper, staining and streaking the surface. Bitumen coating of copper surfaces should not occur and there should be no possibility of runoff of water from bitumen products. There is a proprietory system which uses bitumen-backed copper for roofing which is acceptable as it is shielded from rainwater. Red cedar wood also produces acid in solution which can attack copper and the two materials should not be used together, particularly if there is the likelihood of water runoff from the cedar wood, often used as shingles or shakes.

All copper alloys are susceptible to attack by ammonia, resulting in a bright blue corrosion product, and assisting in stress corrosion cracking.

Brass

Brass covers a wide range of copper–zinc alloys where copper content will vary from 58 to 95%. They are single phase solutions with solid solutions of zinc in copper. There are approximately 40 standard alloys which can be cast or used to form:

- Extruded rods and sections (hollow and solid).
- Hot stampings and forgings.
- Rolled plate, sheet strip and cut circles.
- Drawn tubes (round and shaped).
- Wire, round and shaped castings, sand, shell, investment, gravity and pressure die.

Figure 8.4 This door in Salzburg, Austria, is made from patches of copper held on by iron rivets. There is some sacrificial corrosion of the iron and local staining, but this seems acceptable in this situation which is already joining materials in an unusual way.

The cold working of brass through drawing or rolling will increase stiffness and strength.

Brass will initially oxidize and produce a tarnish film before developing, over time, a patina similar to that on copper, and the familiar green patina will develop in moist conditions. As the zinc content of the alloys increases, the colour changes from dark red through to strong yellow colours. Lacquering will preserve the colour although the alloys can be enamelled or plated with chromium, nickel, silver or gold. Copper is a known biocide and particularly appropriate for clean environments, such as hospitals, in the choice of fittings.

Alpha brasses contain a minimum of 63% copper and commonly have a copper:zinc ratio of 70:30. This alloy has the best corrosion resistance due to the high copper content, and is ductile with the advantage of added strength gained from cold working. It is the common alloy for wood screws, electric fittings and pressings. An alpha structure is one in which the amount of zinc that can stay in a solid solution with copper at room temperature reaches a maximum at 39%. A beta structure is an intermediate phase of copper–zinc.

Duplex brasses have a higher zinc content and are suitable for extruded rods and bars and also hot stampings for pipe fittings, taps, valves, and window and door furniture. Sections can be easily machined to provide threaded connections for other components. Certain supply water, including sea water, can cause *dezincification*, which is the actual removal of zinc from the alloy leaving behind a porous copper artefact.

Additional metals may affect properties. These include tin, aluminium, manganese and nickel to improve strength, hardness and overall durability. Tin, arsenic, aluminium and nickel also help improve corrosion resistance.

High tensile strength brasses These are higher strength brasses used for machinery components which have extra wearing capabilities. The window frames for Coventry Cathedral are made from these alloys with a natural and described *chocolate brown patina*.

Bronzes

These are alloys of tin and copper and may contain up to 10% tin, staying as single phase material. They are more commonly used in casting.

8.3 Lead

The Lead Development Association, 42 Weymouth Street, London W1N 3LQ, publishes clear manuals for the specification and detailing of lead, including *Lead sheet in building* (reprinted 1989) and *The lead sheet manual*, vol. 1, 1990, which is the first in a series.

Lead is used in its pure form (99.9% purity) for roofing, flashings and damp-proof courses due to its durability and its malleability as a pure metal which allows for complex shaping, vital in angled profiles and bends in roofing. It is also ideal for weather protection on near horizontal stone

Table 8.11 Standard properties of lead

Property	Value
Atomic weight	207.2
Atomic number	82
Density	11.34 g/cm^3
Coefficient of linear expansion	29.7×10^{-6}/K
Thermal conductivity	34.76 W/m K
Melting Point	327.4 °C
Tensile strength *with 1%Sb, cold rolled*	29.8 MN/m^2

Figure 8.5 Common zinc roof details in Salzburg, Austria, with standing seams.

surfaces and, because of its malleability, ideal for detailing around the end grain of exposed timber details. As it is a soft metal it can be used as a load bearing pad, and will deform to give a uniform bearing between materials. It will avoid eccentric or point loading due to the surface roughness of materials (such as concrete) and optimize surface contact for load transference. There is also a useful application in expansion joints for thermal movement, although in engineering structures today other methods are used. It can be effective as insulation against sound because of its density, and can be useful as part of a composite panelling with plywood to assist sound reduction.

Applications

Lead is also used in radiation shielding, and can be incorporated in walls or used for linings to containers in hospitals and by other specialist users. The Lead Development Association has a separate publication relating to levels of protection. Less commonly it is used for forming sinks, particularly where corrosion by chemicals has to be catered for and, less commonly these days, for pool linings and stair treads for resistance to heavy traffic.

It is a bright metal but readily oxidizes with an additional formation of white lead carbonate when newly exposed to moisture. This white carbonate can be washed onto other materials causing staining, especially to darker materials such as slate, but it can be prevented by applying patination oil, originally linseed oil, but now a general *weathering oil* to the metal. If patination oil is not used initially, lead surfaces can be cleaned by using a 5% nitric acid solution, washed with clean water, dried and treated with patination oil. Long term exposure builds up a patina through reaction with carbon dioxide and sulphur dioxide in the atmosphere, producing chiefly a dark lead sulphate as the final patina.

Lead is vulnerable to attack from organic acids which may leach out of particular hardwoods or cedar roofing shingles. Organic acids can also exude from lichen or mosses with water trickling down the lead from either light rain or heavy dew, eventually forming noticeable grooves. BRE Digest 139 lists toxicants that can be used to remove these growths. They include tribtylin oxide in aqueous formation (hazardous chemical) and the LDA recommends the proprietory product *Murosol 20* by Wycamol. A healthier method of control uses copper strips placed at every tenth course of slates or tiles. Copper salts act as a natural biocide inhibiting growth. If the picturesque effect needs to be retained then lead gutters and flashings can be double lined and these additional layers can be treated as sacrificial elements which can be replaced relatively easily when necessary.

There can also be alkali attack from cement or concrete before free lime undergoes conversion to a carbonate during case hardening. This occurs where lead is built into walling adjacent to mortar which is not exposed directly to carbon dioxide. This reaction is likely to be slow, and free lime will be available. The lead should then be protected with a bitumastic paint. Signs that this kind of reaction is happening is when yellow staining occurs on the lead as the corrosion product leaches out of mortar joints. This kind of preventative measure is not necessary for flashings which are tucked into small chases in brickwork, as the relatively small depth allows penetration by carbon dioxide for carbonation of free lime. However, when setting lead flashings into chases or mortar joints it is recommended to apply masking tape to the lead before pointing. This ensures that the mortar adheres to the underside of the chase and, as there is allowance for the lead to move, cracking in the mortar is prevented.

Lead can also be affected by condensation which assists its conversion into lead carbonate. Condensate liquid is distilled water in which lead will slowly dissolve. In closed roof conditions there is usually insufficient free air to give enough carbon dioxide to form a stable patina of lead carbonate. Roofing areas should be well ventilated just below the lead sheet and boarding (see Technical Note 3 by LDA) and, in general, higher standards of insulation together with good vapour barriers should provide adequate protection. As lead heats up in sunshine, sudden showers can cause pressure differentials that pull rainwater through joints, and this free water can then vaporize in hot periods and condense on the underside of the lead. The LDA has also looked at the traditional felt underlays which can rot in this situation and adhere to the lead preventing thermal movement. They now recommend the use of *non-woven polyester* textiles made by Monomet Ltd or Don and Low Ltd (See Technical Note 8 by LDA). If there is little risk of condensation then building paper underlays to BS 1521 can be used as a separating medium.

Lead corrosion

Lead is stable as metal and falls into the category known as *cathodic* metals in the electrochemical series. It can be in contact with copper, zinc, iron, stainless steel and aluminium without great risk of bimetallic corrosion. However, if lead and aluminium are used in marine environments there is a reaction between lead oxide on the surface of lead, and sodium chloride in salt water, which creates a caustic solution that attacks aluminium. Crevice corrosion in these situations is severe, so this combination is not recommended in this specific environment. Generally all types of metal fixings can be suitable provided there is no moisture in the timber substrates. Tests have been carried out that combine different timbers with their likely timber treatments. The fixings that match the longevity of lead are stainless steel, copper and brass. Galvanized mild steel and aluminium fixings will fail first, and negate the reason for specifying a long life material such as lead. (See Technical Note 4, LDA.)

Lead is mostly used in a sheet form and produced by straightforward rolling, until the correct thicknesses are achieved. It is important to specify the right thickness in applications as it is the densest metal used in building and is able to creep under its own weight. If used on south facing elevations, the length of the sheet should be restricted to cater for movement, and lengths of lead may have to be limited to 1.5 m. A 2 m length of lead in hot conditions could move as much as 2 mm. Sometimes lead is cast into sheets for roofwork or for ornamental work. Sheet is specified by a code which relates directly to the thickness of the lead.

It is important to relate the siting of lead to the pitch of roof, or where there may be lengths that have less support and can creep under their own weight (as in valley gutters).

Lead sheet categorization

The BS 1178 categorization of lead sheet is given in Table 8.12.

Fixings

Fixings should be copper clout nails to table 2 BS 1202 Part 2, not less than 25 mm long or 10 swg. Screws should be of brass or stainless steel and conform to BS 1210.

Table 8.12 Lead sheet categories

Coding		Thickness (mm)	Weight (kg/m^2)
Number	Colour		
3	green	1.32	14.97
4	blue	1.8	20.41
5	red	2.24	25.4
6	black	2.65	30.05
7	white	3.15	35.72
8	orange	3.55	40.26

As lead is expensive and heavy, an alternative is to use laminated sheet materials such as plywood, chipboard or steel sheet, which can be bonded to a coating of lead which may be as little as 1 mm in thickness. Adhesives are used for bonding to board materials but lead can be satisfactorily bonded to steel using a lead tin–alloy, giving an intimate metallurgical bonding.

Lead Working

Lead is worked and joined in two main ways. It can be *bossed*, meaning the pushing of lead from one area to another, traditionally using hardwood tools, or it can be welded (leadburned). Overbossing can result in an unacceptable thinning of the lead, but can work the material around complex shapes. Lead welding uses a lead solder with an oxyacetylene flame but the resultant joint is harder and more likely to show fatigue cracking. The solder also has a different appearance with a colour variation.

Cleaning of lead

In restoration work some lead may be heavily coated with a black patina. Cleaning gel is a product that can be used to match the old leadwork next to newly installed leadwork. It is described as neither strongly alkaline nor acidic.

Lead and health

The Control of Lead at Work Regulations came into force in 1980. Dependent on the degree of exposure, clothing needs, air monitoring and medical surveillance are specified. Ordinary use of new lead is not thought to provide great risk and handwashing before eating, drinking or smoking (heat vaporizes lead) and leaving work, is adequate precaution. There is a far greater risk in the inhalation of lead dust in removing old sheeting, which may have deteriorated due to dissolution by condensation, and respirators should be used on site when handling this material.

8.4 Zinc

Zinc Development Association, 42 Weymouth Street, London W1 3LQ 0171-499-6636. Zinc is not only used in building for roofing, cladding and flashing but has a major role in the protection of steel, chiefly through galvanizing. World demand for zinc runs consistently at 5 million tonnes per annum. It is a bright metal that develops a matt grey appearance after exposure to the atmosphere, with the formation of zinc carbonate. Pure zinc is very brittle and is usually alloyed with either copper or lead to provide a more malleable metal for roofing and cladding details. For roofing generally it should comply to BS 6561:1985 Parts A and B. Part A outlines zinc–titanium–copper alloys and Part B softer zinc–lead alloys which are used for flashings. Zinc can be laid to falls as low as 2.5° although, ideally, falls should be above 3°. Standard thicknesses vary from 0.65 to 0.8 mm. In ideal conditions roofing can have a long life of approximately 100 years.

In Europe there is a practice of using longer sheets of metal strip generally for roofing and special tools have been developed which fold and lap long standing seams. There is also a craft-based tradition which uses zinc in a way that is totally imitative of traditional stone detailing with extremely complex shapes, including full mouldings and ornamental friezes. One example is the restoration of domes for a post office in Konstanz (Germany) where 99 year old details were totally renewed.

Zinc lies at the anodic end of the electrochemical series and should not be used in contact with copper. Water draining from copper roofs or fittings should not discharge onto zinc although there is no problem from small fixings such as nails, etc. It can be fixed to steel, aluminium and lead without significant corrosion but water running off from these metals onto zinc can cause some staining. Like lead it is not directly affected by Portland cements or mortar but, if built into a wall, should have a protective coating of bitumen paint. It is generally laid over softwood boarding which presents few problems even if the surface has been impregnated with preservatives or fire retardent solutions. Like other metals it is vulnerable to attack from the weak acids that are present in western red cedar or hardwoods such as oak or sweet chestnut.

Table 8.13 Standard properties of zinc

Property		Value
Atomic weight		65.37
Atomic number		30
Density		2700 kg/m^3
Coefficient of linear expansion		27×10^{-6}/K
Thermal conductivity	*alloy A*	113 W/m K
	alloy B	109 W/m K
Melting point	*alloy A*	387 °C
	alloy B	388 °C
Tensile strength	*alloy A*	275 MN/m^2
	alloy B	314 MN/m^2

Alloy A: 39.43 Al + Mg 0.04–05 (nil Cu)
Alloy B: 0.04– 0.5 Mg + 0.75 – 1.25 Cu

Forming

Zinc can be formed by hand or with special tools and can be soldered to provide weather-tight joints, with solders conforming to BS 219. Fixings should be of heavily galvanized steel or stainless steel.

Protection of Steel

Zinc provides effective protection by giving a durable weathering coat which, if damaged in any way, provides a sacrificial coating to steel by corroding preferentially to the underlying steel substrate. Zinc can be applied by brush, spray or by dipping on site or under controlled factory conditions. Hot dipped coatings are referred to as *galvanizing*. The method of application and depth of protection has to be determined by the performance required. Factory-controlled conditions are preferred for maximizing protection, as dipping will give a consistent coating to all parts of an item, whereas painting inevitably gives a thinning of surface film over corners and other arrises. The coatings are effective because they adhere to steel by forming an intermediate metallic layer.

Increased protection inevitably has a cost implication. The Galvanisers Association publishes a *Steelwork protection guide* which gives a full breakdown of protection available. BS 5493:1984 *Code of practice for protection of iron and steel against corrosion* and the Zinc Development Association in conjunction with British Steel also publish Steelwork corrosion protection guides. Internal protection is just as important as external protection and the corrosion of steel is possible from condensation. In these situations zinc can prolong the life of steel by 40 years. Zinc-coated steel immersed in cold water distribution systems is exposed to the buildup of salts on the galvanized layer which provides protection of a further 40 years. In hot water systems this additional life is reduced to 10 years. The worst conditions would be in buried soil, and the life of the protected steel would then vary between 5 and 20 years depending on the pH of the soil. See also *MBS*: *Finishes* Chapter 2 Polymeric materials, where effectiveness of coatings is explained in relation to the initial preparation of metals.

Galvanized coatings can be protected by paint finishes but there are problems in obtaining good adhesion, and pre-treatments should be used to prepare the surface prior to painting.

Galvanized steel can be welded but this should not be

carried out in confined spaces as zinc oxide fumes can give influenza-like symptoms. Coatings that are damaged as a result should be repaired by the following (in order of effectiveness):

(1) Grit blasting and spraying to 100 micron thick zinc coating.
(2) Wire brushed, blow torch heated to 300 °C and molten zinc rod or powder applied.
(3) Wire brushed and zinc-rich paint applied four or five coats at 20 micron thickness.

Reference

1 Donald R Askeland 1984 *The science of engineering materials* PWS, Boston.

IV POLYMERS

9 Introduction

In the construction industry there is an important group of organic-based engineering materials commonly known as plastics which are all polymers, a chemical classification for compounds characterized by long chain molecules. When individual molecules or *mers* are linked together they will form these long chain which can be several thousands of units in length. Although polymers exist in nature most polymeric products used in the building industry are synthetic so their exact configuration and composition can be engineered according to use. Products can range from rigid pipes and sheets to flexible rubbers, adhesives, sealants and coatings. The range of manufacturing processes is immense and components can be cast, moulded, extruded and coated. Many of the processes are dependent on the viscosity of the polymer or their ability to flow. Polymers are usually combined with other compounds which act as stabilizers and fillers; the latter help to reduce the cost of an otherwise expensive product. Their combination with sheet forms of woven cloth, fibres or other particles will affect their structural behaviour and they will then become true composites with a considerable increase in stiffness and tensile strength.

It is important to realize that the origin of polymers is in the finite resources of crude oil, gas and coal. Chlorine and nitrogen are gases added to yield the variety of primary monomers needed. Controlled reactions using heat, pressure and catalysts (which can be metallic compounds or acids) yield a variety of monomers which at a later stage are processed to link and form the long molecular chains or polymers. There are now developments to form ethylene, one of the most important monomers, directly from live plant products, an immediately renewable and manageable resource. The technologies for this type of processing were largely developed for third world countries which do not have mineral resources. As oil and gas reserves diminish this becomes a growth area for development in first world countries with a shift in farming priorities. The major

Figure 9.1 Display of polymer resins from the science exhibition of Parc Villette showing the raw state of polymers prior to processing.

monomers produced from primary resources which yield all the major polymer groups are:

- Ethylene
- Propylene
- Styrene
- Vinyl chloride
- Butadiene
- Cyclohexane
- Acetylene

9.1 Polymerization

Polymerization as a process is the engineering of controlled chain growth and chemical structure. The activity for the design of new polymers is often referred to as *polymer architecture*. The basic monomer (individual monomer) or building block has to be altered to allow it to combine with adjacent molecules to make a polymer or long chain molecule. This involves breaking chemical bonds by using radicals, highly reactive molecules or atoms which have an odd number of electrons. They attack the monomers, often breaking double carbon bonds yielding unpaired electrons which then have the ability to form links. Chain length has to be controlled. This can happen by radicals joining together or becoming stable by altering their own chemical structure by reacting with each other and then being unable to initiate further reactions, or by reacting in such a way as to produce side branching on a chain still with effective termination of the chain length. Chain length is particularly critical in rubber production where as chain length increases the mixture becomes too viscous to be processed properly. The chemistry of the polymerization process is complex and is oversimplified here, but it is important that the concept of long chain molecules is understood as this leads to greater understanding of their properties and subsequent degradation.

Polymers are divided into two major groups, *thermoplastics* and *thermosets*. The group known as *elastomers* have additional properties which show extensibility and include rubbers and sealants but even this category subdivides into those that are engineered as thermosets or thermoplastics. Most polymers we use were engineered between the 1930s and the 1950s. Phenol formaldehyde (known as Bakelite) was developed in 1909 by Leo Bakeland.

Plastics used in building are found most commonly in the following areas:

Services Water supply, gas supply drainage, ventilation or electrical wiring and conduits.

Accessories for damp-proofing Membranes and damp-proof courses, water stops.

Thermal insulation Boards, foam filling of cavities, pipe insulation.

Acoustic insulation Impact sound absorption, acoustic absorption, isolators for vibration.

Daylighting control Windows, roof-light frames.

Cladding Roofing and wall panels, boarding.

Components Doors, hinges, frames, door furniture, flooring, tiles and sheets.

Composite materials In combination with wood products for boards, lightweight aggregates for concrete.

Coatings and finishes Paints, sealants, protective film coatings that are bonded with adhesives. Films for solar control on glass. (See *MBS: Finishes* Chapter 2 (Paint technology).)

Sheet material/membranes Roofing and flooring, air-supported structures.

The principal polymers in use are shown in Table 9.1.

Table 9.1 Principal polymers

Type	Name	Abbreviation
Thermoplastics	Polyethylene	PE
	Polypropylene	PP
	Polystyrene	PS
	Poly(vinyl chloride)	PVC
	Polyacetal	POM
	Acrylic	PMMA
	Polyamide (nylon)	PA
	Polycarbonate	PC
	Polytetrafluorethylene	PTFE
Thermosets	Epoxy	EP
	Melamine-formaldehyde	MF
	Urea-formaldehyde	UF
	Unsaturated polyester	UP
	Phenolic	PF
	Alkyd	—
	Polyurethane	PUR
Elastomers	Natural rubber	NR
	Styrene–butadiene rubber	SBR
	Polybutadiene	BR
	Butyl rubber	BUTYL
	Polychloroprene	CR
	Synthetic polyisoprene	IR
	Nitrile	NBR
	Silicone rubber	—

Source: Table 1.1 C Hall 1981 *Polymer materials* Macmillan

9.2 Types and processing of polymers

Thermoplastics

These polymers are the largest group of polymeric materials and have the characteristic of individual long chain molecules which can slide past each other. These chains may have side groups which stiffen them, giving them different properties. Sometimes local alignment of these chains is so regular, they have sufficient local order to be referred to as crystalline polymers.

This group has the greatest application in industry due to the ability of thermoplastics to soften on reheating and to be reformed. This allows for their processing to yield a variety of products which can utilize a wide range of moulding and extrusion techniques.

Polyethylene PE (high and low density, HDPE and LDPE)
Polypropylene PP
Polystyrene PS (toughened HIPS)
Poly(vinylchloride) PVC
Polyacetal POM
Polymethylmethacrylate (acrylic) PMMA
Polyamide (nylon) PA
Polycarbonate PC
Polytetrafluorethylene PTFE

Figure 9.2 Early bakelite radio and turntable.

Polyester–urethane PU
Cellulose acetate CA
Acrylonitrile–butadiene–styrene ABS
Styrene–acrylonitrile copolymer SAN
Poly(vinyl butyral) PVB
Poly(vinyl fluoride) PVF

By fast cooling liquid thermoplastics, their structure can be made as strong as some thermosets. With the addition of carbon fibre reinforcement, components can be made from poly(ether–ether–ketone) (PEEK) that replace parts in aircraft traditionally made from metal. Silica may be added to acrylic resins to provide the hard marble-like surfaces used in washrooms or kitchens. These kinds of innovation bring into the world of engineering plastics which account for approximately one tenth of all the plastics produced. They include the following polymers:

Polypropylene
Polyamide (nylon)
Polycarbonate
Thermoset epoxy resins
Polyacetyl
Other thermoplastics

Most of these are easy to recycle as they are predominantly thermoplastics and have a known composition.

Thermosets

The chief characteristics of these long chain molecules is that they are linked sideways to each other, referred to as *cross-linking*. This gives greater rigidity to the material and any movement is incremental, sometimes shown by brittle behaviour. Once made, these polymers will set irreversibly and the final reaction in processing has to be carefully controlled. The two-pack preparations used on site for adhesives and coatings are usually thermosets. Great heat is usually generated in the final chemical reaction. The application of thermosets is more specific and they are not easy to recycle.

Epoxy EP
Melamine–formaldehyde MF
Urea–formaldehyde UF
Unsaturated polyester UP
Phenolic-formaldehyde PF
Alkyd
Polyurethane PUR

Elastomers

The carbon bonds in elastomers are able to rotate which allows considerable movement under stress, which is reversible once the load is removed. This phenomena gives the characteristic elastic behaviour where extensibility of the polymer can sometimes be as much as 300%. Most elastomers or rubbers can be recycled for rubber crumb in car tyre production and playing surfaces.

Natural rubber NR
Styrene–butadiene rubber SBR
Polybutadiene BR
Butyl rubber BUTYL
Polychloropene CR
Synthetic polyisoprene IR
Nitrile NBR
Silicone rubber

Polymer architecture will adjust the chemistry of the polymers and often combine polymers to optimize performance. Synthetic rubbers are elastomers but are cross-linked to provide stiffness and durability.

Elastomers are the largest group in terms of actual production and the tonnage produced is four times that of thermoplastics due to the tyre industry. Thermosets form only one-fifth of the tonnage for thermoplastics.

Manufacturing processes

Although plastics may be made from the same resin, they can be changed totally by additives used and by the method of manufacturing. This can be divided broadly into the following groups.

Moulding There are several types of moulding which include *compression moulding* which heats the polymer under pressure in a preformed shaped which is then cooled, *transfer moulding* which uses pressure to make the polymer flow into the preformed shape which is kept under pressure until it has set and cooled, and *injection moulding* which uses a higher pressure to transfer the polymer into the mould and generally speeds up production.

Extrusion In this process polymers are fed under pressure usually being forced through a die by the turning of an extruder mixer screw which forces material along a tube to the preformed die. Under pressure the material is then pushed into a preformed shape in much the same way as toothpaste is squeezed from a tube. As the material leaves the die, lateral pressure is released and the polymer may increase in size, which is known as *die-swell*. Hollow articles can also be made by extrusion blow moulding where the polymer is inflated into a reverse mould shape, using a similar technique to glass blowing. This allows for the fast repetitive production of objects. Profile extrusion produces more complex sections with greater accuracy using polymers other than polyethylene.

Flat sheets (over 0.25 mm thick) are produced from a die in the form of a long slit. It has to be horizontally extruded as self-weight would lead to deformation of the material.

Sheet material (under 0.25 mm thick) is referred to as film and the molten material from the die usually flows vertically over chilled rollers in a vertical cooling tower. Alternatively tubes of film can be made by swelling a cylinder of molten polymer with compressed air which is then cooled, flattened between rollers and then wound.

Calendering Sheets can also be produced by calendering which involves the molten polymer being squeezed progressively between rollers until it is the required thickness. The surface of the rollers will also determine the finished surface quality of the sheet, and can be imprinted with textures and finished matt or polished.

Pultrusion This term is used more widely in the USA and describes a process restricted to patent holders. It allows for the continuous pulling of fibres through catalyzed resin baths and then through heated steel dies which start to cure the resin. The end product will have great directional strength. The technique of cold drawing aligns crystalline regions in a polymer. The increase in strength is so considerable that polymeric binding tape produced by this method can replace steel tape. This technique is also used with continuous mats or fabrics where a composite material is built up as a series of plies and can be pultruded as a complete composite, flat, round or square in section. For thick sections, normally regarded as being over 3 mm in depth, the incorporation of a centre mat ply will help to arrest cracking.

Generally ply construction should be placed symmetrically around a central axis. The glass fibre content in these composites can be as much as 50%, typically used in conjunction with a polyester resin. The distribution of the mat reinforcement must respond to the type of loading on the element. If there is torsional loading then there must be a circumferentially wound system for the fibre reinforcement. If mats or reinforcement are discontinuous there must be an overlap of at least 12 mm which is then likely to give at least 80% of the strength of the section. These systems of composite engineering are used not just for sheet materials but for round and square section tubes.[1]

Extrusion coating, laminating and melt coating are similar processes that combine other materials with polymers. The pultrusion products have greater structural strength and more relevance in the building industry.

Casting Polymers can be cast into shape by completing the polymerization process (which may use a resin and hardener) within a mould. Alternatively a process similar to slip casting in pottery can be used. Here, a fine slurry of polymer powders is carried in a liquid medium coating the inside of a heated mould which is then fused to form, for example, hollow spherical shapes. Powder casting techniques use the polymer directly without any medium and are particularly useful for coatings.

Thermoforming vacuum forming This process uses thermoplastics which are laid in sheet form on top of a mould, heated to deform into a loose fit following the mould profile. A vacuum is then applied to pull the sheet into its final shape. If air pressure is applied as well for heavier materials the process is known as *pressure forming* and more complex shapes can also be made. Once the object is made the vacuum is released and the finished object pulled away.

Typically the plastic waffles or troughs used as formwork for casting slabs would be made in this way.

This process allows for the formation of complex shapes. To understand the stresses involved in making these shapes, a demonstration using sweet or biscuit trays will show how this process can be reversed. Put these objects in an oven at 100–150 °C for ten minutes (taking the precaution of laying them on aluminium foil on a baking tray). With the re-application of heat, stress is released from the polymer that was applied in the original processing to cause deformation, and the material flows back to a rigid sheet form, usually much reduced in size. This demonstrates the phenomenon of plastic memory, and shows how the manufacturing process relies on the viscous flow of the material, and that the retention of the shape required is also temperature dependent.

Cellular polymers/foams Voids are formed in the polymer to develop rigid or flexible cellular materials. They can be regarded as composites of solid and gas. Air can be introduced mechanically through stirring in a liquid phase; lowering the pressure can also produce cavities and chemical reactions can produce gas bubbles. These mechanisms help in the production of insulant materials and can also produce strong structural foams which can then be backed with metal or solid polymer sheet materials to form panels, which compensate for any reduction in tensile strength. Foamed rubbers find extensive use in furniture making, as compressible sealants in building and as components that give sound reduction and sound dampening systems.

Liquid resins Polymer resins may be modified and combined with solvents, extenders and pigments for paint systems. A fuller description of the technology needed for coating systems is described in *MBS*: *Finishes*. Resins are also important as the matrix component in composites that use wood waste to make fibreboards and medium density fibreboards as well as bonding sheet materials such as plywoods. They also form the bulk of adhesives produced. For the technology of adhesion see also *MBS*: *Finishes*.

Reference

1 Joseph E Sumerack and Jeffrey D Martin 1988 *Property of tubular pultrusions via materials orientation and processing selection*. Paper presented at the Annual Conference, Composite Institute, The Society of the Plastics Industry Inc (USA), 1–5 February.

10 Polymers

10.1 Specification and identification

The specification of polymers relates directly to their properties. As the numbers of polymers are so large it is worthwhile to draw up a checklist of desirable properties needed when specifying a polymer, arriving at a performance specification. Properties to be considered are:

Elastic modulus (Young's modulus)
Rubber elasticity
Viscoelasticity
Creep
Response to impact loading
Fracture characteristics
Friction
Abrasion resistance
Response to fire
Softening point
Electrical properties
Optical properties
Thermal performance
Environmental stress cracking
Chemical resistance
Biological resistance
Permeability
Photo-oxidation and UV stability
Radiation

Inevitably no single polymer will be able to optimize on all of these properties and copolymers may be needed, or composites which selectively use the performance of all materials to fulfil a performance specification. See Table 10.1.

Basic polymeric materials can be modified to improve their performance by their incorporation as copolymers or by additives that have the following modifying results:

Antimicrobials
Antistats
Blowing agents
Colorants
Fillers
Flame retardants
Impact modifiers
Lubricants
Plasticizers
Reinforcements
Slip and antiblock agents
Stabilizers

Identification of plastics

Although some resins have characteristic original colours because of the great range of modifiers used, it would very difficult to determine the actual polymer used for a solid plastic solely by colour identification.

The *Identification of plastics* kit produced by Building Research Station gives a number of simple tests that can be used to identify plastics. Pages 7–8 give the basic characteristics of the main polymers.

Transparent plastics can include the following polymers which will give varying degrees of transparency:

Glass clear
Polymethylmethacrylate
Polycarbonate
Unplasticized PVC
Polystyrene

Transparent
Polyethylene
Polypropylene
Polyesters (with or without glass fibre reinforcement)
Plasticized PVC

Table 10.1 Characteristics of common polymers

Plastics \ Characteristics	High impact strength	High scratch resistance	Good weatherability	Readily electroplated	High tensile strength	Resists acids	Resists alkalis and solvents	Good colorability	Low friction	Excellent optical properties	Low moisture absorption	Good adhesives	Good electrical properties	Arc and track resistant	Good corrosion resistances	High temperature capability	Low density	Large volume, low cost usage
Acrylonitrile butadiene styrene	•	•	•	•				•										
Acetals	•				•		•	•	•									
Acrylics						•	•	•		•	•	•						
Alkyds											•		•	•				
Cellulosics																		
Cellulose acetate			•					•		•								
Cellulose acetate butyrate															•			
Cellulose acetate propionate	•									•								
Cellulose nitrate			•					•										
Ethyl cellulose	•										•							
Epoxies		•			•	•							•		•			
Fluorocarbons									•				•		•	•		
Melamines		•						•			•	•	•	•				
Nylons									•			•				•		
Phenolics					•						•	•	•	•	•	•		
Polycarbonates								•							•	•		
Polyesters	•		•			•					•		•		•			
Polyethylene																		•
Polyimide					•	•			•			•				•		
Polypropylene						•	•	•								•		
Polystyrene						•	•	•			•				•		•	
Polyvinyls																		
Polyvinyl acetate												•						
Polyvinyl butyral										•	•	•						
Polyvinyl chloride																		•
Vinyl copolymers		•																
Silicones						•			•				•	•	•	•	•	
Ureas		•				•	•					•						

Source: Table 9 D 1980 P Hanley *Introduction to the selection of engineering materials* Van Nostrand Reinhold

Table 10.2 Typical properties of plastics used in building

Material	Density (kg/m³)	Coefficient of linear expansion		Max. temperature recommended for continuous operation (°C)	Short-term tensile strength (MN/m²)	Behaviour in fire
		$\times 10^{-5}$/K	mm/m			
Polythene* low density	910	20	0·2	80	7–16	Melts and burns like paraffin wax
high density	945	14	0·14	104	20–38	
Polypropylene	900	11	0·11	120	34	Melts and burns like paraffin wax
Polymethyl methacrylate (acrylic)	1185	7	0·07	80	70	Melts and burns readily
Rigid PVC (UPVC)	1395	5	0·05	65	55	Melts but burns only with great difficulty
Post-chlorinated PVC (CPVC)	1300–1500	7	0·07	100	55	Melts but burns only with great difficulty
Plasticised PVC	1280	7	0·07	40–65	10–24	Melts, may burn, depending on plasticiser used
Acetal resin	1410	8	0·08	80	62	Softens and burns readily
ABS	1060	7	0·07	90	40	Melts and burns readily
Nylon	1120	8	0·08	80–120	50–80	Melts, burns with difficulty
Polycarbonate	1200	7	0·07	110	55–70	Melts, burns with difficulty
Phenolic laminates	1410	3	0·03	120	80	Highly resistant to ignition
GRP laminates	1600	2	0·02	90–150	100	Usually inflammable. Relatively flame-retardant grades are available

Source: Table 1 BRE Digest 69

10.2 Durability

BRE Digest 69, *Durability and application of plastics*, 1977, still gives valid parameters for durability which are grouped under sunlight, heat and moisture. Environmental stress cracking is accepted as a term for embrittlement that is caused by the use of abrasive cleaners and chiefly organic chemical agents and gases. It can be accelerated by repeated loading, the best examples of which can be seen close up in domestic examples like washing-up bowls rather than in building components. The same kind of damage can arise in services conduits and this can be unsightly and will ultimately lead to the propagation of cracks.

Generally ageing and long term durability will be dependent on the polymer's ability to remain stable under the long-term exposure in a particular environment. A range of factors will ultimately lead to the breakdown of the polymer. This will include the macro effects of particle-carrying wind, rain and sun as well as degradation from the factors described in Table 10.2.

Degradation by ultraviolet light and oxidation

It is the energy obtained from ultraviolet light that alters the chemical bonding of polymers by causing chain scission, oxidation, cross-linking, hardening and subsequent embrittlement. Small hairline cracks will also give an uneven reflection of light back from the material surface, and this leads to an apparent colour loss. The degradation process involves chemical reactions which increase in activity with rises in temperature, and moisture may also be needed for reactions to occur. Ironically, polymers may be in a more protected situation in urban areas where the effects of ultraviolet light are weakened by protective deposits of dirt and less intensity of light due to micro-climatic conditions. Polymers likely to suffer deterioration under ultraviolet radiation should have added pigments (usually black or other dark colours) that can act as absorbers or should be finished white to reflect harmful rays. Ultraviolet stabilizers are chemicals which can be added to the polymer, change under the effect of the energy from exposure and then revert to their original configuration once the source of ultraviolet light is removed.

Effects of heat and fire

Most polymers degrade rapidly under the effects of heat; polymer chains are broken, free radicals are formed, noxious gases may be released and there will be severe discoloration and deformation. At temperatures above 400 °C deterioration is more dramatic, side groups are split away from the main carbon backbone and charring results, often with a stable ring structure incorporating nitrogen as well as carbon atoms. Sometimes this charring is useful in preventing further deterioration. The burning of timber is well known in producing a protective medium and this mechanism is used in the choice of nylon–phenolic polymers employed as heat shields for the re-entry of space vehicles where the surface ablates, degrading in progressive layers but giving protection which will exceed several thousand degrees.

The flexible foam polyurethane will degrade to yield carbon monoxide hydrogen cyanide and toxic nitrogenous products. These gases and the dense smoke formed are extremely dangerous in fire and a major cause of loss of life.

Most thermoplastics start to soften between 75 and 150 °C. Their flash ignition temperature can be as low as 290 °C and probably not more than 500 °C. This makes them extremely vulnerable in fire which can raise ambient temperatures to over 600 °C in a domestic room. At this point all polymers can ignite spontaneously with *flashover* conditions.

Public concern about fire and plastics was first aroused with the Summerland fire disaster of 1973 on the Isle of Man where pyramidal PMMA panels clad the roof and upper wall area. The roof panels in collapse did help to vent the fire and, despite their poor performance, these alone were not singled out as the cause of the tragedy which arose from a multiplicity of factors.

Depending on the way plastics have been processed, they will be formed into shapes after the application of heat and then cooled, so they will be in a state where stress is frozen into the artefact. If that stress is released through reheating, there can be dramatic changes in shape. The PMMA sheets used on the Munich Olympic Stadium had been made by a method which drew the PMMA in one direction. If heated, stress relieving took place at only 100 °C and the sheets shortened and fell out.[1]

Detailing then has to allow for this kind of phenomenon so that panels cannot fall. Thermosetting products are more stable and may char but still prevent fire penetration to other materials or areas. Polymethylmethacrylate (PMMA), polycarbonate (PC), acrylonitrile–butadiene–styrene (ABS), polystyrene (PS), polyethylene (PE), polypropylene (PP), glass-reinforced polymer (GRP) and expanded polystyrene (EPS) all generate carbon monoxide. Poly(vinylchloride) (PVC) generates chiefly hydrogen chloride as well as carbon monoxide. Most deaths in fires are caused by the effects of smoke and poisoning by carbon monoxide generated by the incomplete combustion of materials. However, there are also a small number of deaths owing to nitrogen-containing substances, such as polyurethane rubber (PUR), which can also generate hydrogen cyanide in a limited air supply, as well as carbon monoxide.

Although perspex has been used to clad complex roof structures, a system now developed in Germany uses ordinary glass. Such a system was used to cover the courtyard of the History of Hamburg Museum by the architects von Gerkan, Marg and Partner in 1990. Solid galvanized steel roofing bars, finished with a white polyester coating, are bolted together to form a curved lattice stiffened internally with tension cables. The glass used is 12 mm thick safety glass with neoprene seals and gaskets. It is strong enough to walk on for maintenance and to carry the heavy snow loads expected in this region. (See *The Architects' Journal*, Focus, April 1991.)

A general point from BRE Digest 22, *Toxic products of combustion of plastics*, 1976, notes that materials treated initially for flammability had used chemicals that would increase the amount of smoke and toxic gases involved in a fire.

Permeability

Gases and vapours can dissolve through polymers and this is a more critical problem if polymers are used as thin sheet materials. This phenomenon increases with temperature and is pressure dependent. It may be necessary to specify polymers which have low permeabilities in specific situations and these will include polyvinylidene chloride (PVDC) and copolymers, acrylonitrile–styrene copolymers, epoxy resins, polyvinylidene fluoride (PVDF), polyethylene tetraphthalate (PETP) and PVC.

Water vapour resistance is the reciprocal of the water vapour permeance of membranes.

Water absorption

It is important that polymers are also used in situations where their absorption of water is not a problem. If moisture is present it can help the de-bonding of composites by weakening the hydrogen bonding mechanisms between long chain molecules, layers or fibres. Moisture can be absorbed by some polymers which swell to cause local stiffening, the material cannot then flex under stress and subsequently fractures. Some nylons have such a high level of water absorption that the resulting dimensional change may then severely affect their performance. Components that rely on having a tight fit may stick or may transfer stress to other materials and cause cracking, particularly in glass.

The unusual juxtaposition of components that may affect polymers, particularly in damp conditions, should be checked with manufacturers.

Biological attack

When polymers are attacked by micro-organisms it is more likely to be as a result of incorporated plasticizers rather than the actual polymer, although plasticized PVC is generally recognized to be one of the most vulnerable. Adhesives and paint films that incorporate natural oils, as well as casein or cellulose are likely to suffer attack.

Chemical attack

Solvents do not completey dissolve polymers; once the solvent has evaporated, the main polymer chain usually remains unaffected although it may physically appear to be in some disarray. Most polymers are fairly resistant to chemicals. More damage is done to primary polymer chains by the effect of ozone which breaks double carbon bonds, decreasing chain length and causing embrittlement. Rubbers are especially vulnerable.

The choice of polymers is inevitably decided by their end use and the polymers listed in the following section on insulation and cellular polymers are used to make products which are controlled by British Standards. These legislate as to the minimum properties required by the materials for those situations. An additional criterion would now be the use of CFC-free materials in the interests of minimizing the use of materials that deplete the ozone layer as CFCs have been used traditionally as foaming agents or blowing agents. Legislation has not yet formalized globally-agreed policy and targets, e.g. the Rio Agreement.

10.3 Cellular polymers

Processing changes polymers and their final application and one group of materials known as cellular polymers change their nature by the controlled introduction of voids. The blowing agents used include CFCs, which are used as they give the whole polymer–gas composite better strength, adhesion and waterproofing properties. This has been particularly important in rigid urethane foams (including polyurethane and polyisocyanurate foams, extruded and expanded polystyrene foam). There are alternative gases which include CO_2 and CFCs. The types of cellular polymer can be recognized to some extent by their colour:

Polyurethanes	white grey–grey brown
Polystyrene	white
Polyvinylchloride	yellow–yellow, brown–dark brown
Polyethylene	white–dark grey–black, pastel shades
Polypropylene	colourless
Silicones	colourless

Rigid cellular polymers

These rigid materials are used for decking, tiles and cavity wall insulation. There is a large number of Agrément certificates relating to individual products but the main British Standards are:

BS 3837 *Expanded polystyrene boards*. Part 1:1986 *Specification for boards manufactured from expandable beads*. (See also BS 6203 *Guide to fire characteristics and fire performance of expanded polystyrene EPS used in building applications*.)

Different grades of boards relate to polystyrene with different levels of impact loading and whether a flame retardant is incorporated with a colour coding.

SD: standard duty — yellow Non-load bearing applications, cavity wall insulation, void formers for concrete shuttering, domestic floor insulation.

HD: high duty — black Flat roof insulation for maintenance traffic, sandwich panels, external concrete shuttering, floor insulation.

EHD: extra high duty — green Flat roofs with higher traffic loads especially for balconies and terraces.

UHD: ultra high duty — orange Special applications, e.g. industrial.

ISD: impact sound duty — yellow and violet Used for reduction of transmission of impact sound.

Flame retardant board is indicated by an additional red stripe if cut and an impressed letter A if moulded.

Each grade has a slightly different performance in respect of compressive stress which if increased will give a slightly poorer performance in thermal conductivity and water vapour permeability.

Preferred sizes for the building industry are:

Length 1200, 1220, 2400, 2440 mm
Width 600, 610, 1200, 1220 mm
Thickness 13, 25, 40, 50 75 100 mm

BS 3837 Part 2:1990 *Specification for extruded boards*. Different grades of boards relate to polystyrene with different levels of impact loading and whether a flame retardant is incorporated with a colour coding. The extruded boards have a better performance in terms of lower thermal conductivity. These boards have greater applications as parts of sandwich panels.

E1: Limited in load bearing, cavity wall insulation.
E2: Lightload domestic flooring and warm roofs.
E3: Limited load bearing capabilities in agricultural buildings, pitched roof construction, medium load domestic flooring and warm roofs.
E4: Medium load bearing for flat roofs.
E5: High load bearing for coldstores and car park areas.
E6: Suitable for lamination to rigid sheets such as metal or plasterboard, able to take wet render and plaster finishes.
E7: As E6 but with greater compressive strength and improved thermal conductivity.

Skinned boards have surfaces formed as a result of their extrusion process. Planed boards (E6 and E7 only) have this surface removed. Preferred sizes are:

Length up to 2500 mm
Width up to 610 mm
Thickness up to 110 mm

BS 3927:1986 Rigid phenolic foam PF for thermal insulation in the form of slabs and profiled sections These boards come in three types A, B and C differing with respect to thermal conductivity, apparent water absorption and water vapour permeability. The standard also covers pipe insulation. (See BS 6336 for recommendations with regard to fire.) Lengths over 4000 mm are available. They have a wide temperature tolerance range of between −180 °C and +130 °C.

BS 4841 *Specification for rigid urethane foam for building applications* is in three parts:

Part 1: Laminated board for general purposes This board has very low thermal conductivity at 0.024 W/m K and is suitable for cavity insulation in heavy construction and under screeds, but is not recommended for ceiling or roofing insulation. There is particular concern over its fire spread capability. The boards are usually faced with paper. The sizes are standard at 2400 mm for length and 1200 mm for width, thicknesses vary from 12 mm to 50 mm. If the basic polymer was modified to improve performance in fire concern was also expressed in this standard over the level of smoke and toxic gases likely to be generated.

The physical property requirements of laminated board are given in Table 10.3.

Table 10.3 Laminated board physical property requirements

Physical property	Test requirement
Compressive strength stress of core at 10% deflection (minimum)	140 kPa
Water vapour permeability of foam core at 38 °C and 88% r.h. (maximum)	5·5 ng (Pa s m) PUR boards 8·5 ng (Pa s m) PIR boards
Dimensional stability, linear change (maximum)	
at −15 ± 2 °C	0·5%
at 70 ± 2 °C and 95 ± 5% r.h.	5·0%
Burning characteristics of small specimens of foam core	Extent of burn less than 125 mm

Source: Table 1 BS 4841 Part 1: 1993

Part 2: 1975 Laminated board for use as wall and ceiling insulation Protection is asked for in this standard by the application of 9.5 mm plasterboard or two coats of gypsum plaster. The co-ordinating size is 2400 × 1200 mm.

Part 3: Rigid polyurethane (PUR) and polyisocyanurate (PIR) foam for building applications Specification for two types of laminated board (roof boards) with auto-adhesively bonded reinforcing facings for use as roof board thermal insulation for built up roofs.

The facing materials are either bitumen coated glass reinforcing facings or glass–polyethylene–glass reinforcing facings. These composite panels are likely to be mounted on steel decking systems.

Flexible cellular polymers

Flexible polymers are more likely to have lower-scale uses such as in upholstery for furniture. Polyurethane might be one of the most common. Fire Research Note 1017 is still applicable in the identification of fire hazards from these types of foams which produce smoke and toxic gases of carbon monoxide, hydrogen cyanide and nitrogen oxides at high temperatures and toxic isocyanates at low temperatures.

Cavity insulation

There is increasing awareness as to the shortfalls of filled cavity insulation. Chemicals used in foamed fill, which include the formaldehyde groups, can emit gases which can cause irritation to occupants, leading to a variety of ailments including sore throats and irritation to the eyes and nose. This happens if there are pathways for gases to escape through breaks in the plaster such as around window and door openings, where rooms are partly within roof spaces, and around floor joists which are seated or attached to unplastered walls which will allow gas penetration through mortar joints. Impermeable floor coverings are recommended and it may be advisable to lay hardboard over polythene. Permanent ventilation should also be increased to properties which have had this treatment.

If cracks form in the foam they create capillary networks and pathways which can increase the flow of water from the external wall leaf to the inside and the top of the foam fill will act as a cavity bridge. BRE Information Paper 7/84, March 1984, acknowledges the contribution made by the upgrading of homes (over one million dwellings up to 1984) and the benefits in energy conservation. However, the Paper recognizes the health hazards which can last for years.

The foam is generated by injecting liquid with an aqueous foaming hardener which sets into fine uniform foam after chemical reaction with oxygen in the atmosphere. Urea formaldehyde foam systems are regulated by BS 5617:1985 *UF foam systems suitable for thermal insulation of cavity walls with masonry or concrete inner and outer leaves*. Complete hardening will take up to three weeks and the material produced is strong enough to support its own weight in a cavity up to a maximum height of 12 m. BS 5618:1985, amended March 1990, is the code of practice relating to the installation of these systems and gives a detailed series of points which form the basis of a checklist for installation. The code also relates to standard cavity wall systems of brick and does not apply to random rubble walling. The exposure index of a building is an important factor in deciding whether or not cavity foam can be used as it will help predict likely rain penetration.

Checks on installation to buildings should ensure that there are no gaps or cracks. The foam should stop short if there are independent damp-proof courses to internal and external walls, achieved by spacing injection holes closer together so the foam can form a natural structural bridge.

10.4 Roofing membranes

Although there are a large number of Agrément certificates relating to the use of polymeric materials such as polyester mats, PVC sheeting (sometimes glass-reinforced with plasticized) and rubber sheeting, there are no British standards that cover these materials used in roofing. These certificates refer back to the Building Regulations to show compliance with regard to weather tightness, ability to take traffic load, for maintenance and adequate response to minor structural movement and temperature variations. Reference may also be made to an equivalent performance expected by more traditional sheet roofing systems in relevant codes of practice. Certificates should also show that adhesion of the membrane to any decking or holding down mechanism is adequate to resist wind loadings expected for particular regions or exposure conditions.

Because of the tendency for some polymers to migrate into similar compounds there should be separating layers of fabric or aluminium foil if they are to be used as roofing membranes. This relates to membranes laid in contact with bitumastic finishes, polystyrene boards or timber substrates that have been impregnated with preservatives.

There are also MOAT (method of assessment and testing) directives for the assessment of waterproofing systems; MOAT 29: 1980, for example, is a specific directive for the evaluation of PVC sheeting systems. Refer also to the UEAfc (European Union of Agrément) directive *Assessment of manufactured plastic flooring*.

BS 3012:1970 and 1980 *Specification for low and intermediate density polyethylene sheet for general purposes* specifies sheet with a nominal thickness of not less than 0.5 mm, a density not greater than 0.93 g/ml and a tensile strength over 9.8 MN/m^2. If the sheeting is to be exposed to sunlight it should be black with a carbon content between 2 and 3%.

BS 4646:1970 *Specification for high density polyethylene sheet* relates to polyethylene with a density over 0.93 g/ml and a tensile strength of not less than 19 MN/m^2.

Agreement certificates will relate performance assessments to density water vapour permeability and the effectiveness of jointing techniques, as well as compliance with current Building Regulations.

Fabric technologies for roofs

These structures are becoming more viable in providing elegant lightweight weather protection for unusual situations. They are also ideal for forms with double curvatures which are difficult to make in other materials, although thin stainless steel sheets using the material property of up to 50% ultimate strain can be deformed by deflation and are used in solar concentrators at Riyadh. The use of fabric membranes is spreading to more permanent installations with the development of double skins that might sandwich insulation. Structural forms can also be worked out in models using nylon, and stress analysis can be undertaken using computer-aided design. These maquettes can be validated by wind tunnel testing.

The Imagination Building, completed August 1989, uses a membrane roof structure carrying a loading of 0.5 tonne/m^2 prestressed to 100 kg/m^2. There are two membranes forming a double skin which are likely to need restressing annually and the outer membrane is coated with a lacquer to help prevent degradation by ultraviolet light. It is a poly(vinylchloride) (PVC) coated polyester with a translucency of 15%. No fungicides were integrated with the PVC and some small areas of growth were noticed in the fabric joints (*The Architects' Journal*, 11 April 1990). It has a guaranteed life of 15 years. It is recognized that polyester cloths are dimensionally stable but can degrade faster with time under the action of ultraviolet and may become embrittled.

The architects were Heron Associates, the engineers Buro Happold and the membrane by Landrell Fabric Engineering.

Teflon-coated glass fibre was first used in this country by Terry Farrell in the form of a 0.5 mm sheet of glass fibre fabric for Clifton Nurseries in Convent Garden. The glass fibre cloth was dipped in liquid Teflon (PTFE) and then cured by heat treatment. Individual sheets were joined by welded seams. Teflon glass fibre used in the Don Valley Sports Stadium is expected to give a 25-year life. As PTFE can give off toxic fumes in fire, analysis had to be carried out to show that this would not occur in the stadium and that temperatures would not reach the critical limit. The glass fibre cloth used as the matrix was also prestressed by stretching over bays. Detailed structural work was carried out by City University.

Silicone-coated glass fibre is another alternative composite membrane. It is very durable and more flexible than the Teflon-coated glass fibre and meets Class 1 surface spread of flame under BS 476 Part 7. The patterning of fabrics is designed to compensate for creep. As they rely on fabric strengths, the warp fibres provide directional strength and the warp provide lateral stability and dampening under wind load. In a glass fibre fabric the warp and weft are loaded differently to compensate for greater shrinkage along the warp and extension along the weft. This allows for the crimping effect characteristic of using cloth structures.

The design of fabric joints and connections has to take into account the behaviour of a fabric under conditions which increase local stress, and ideally the warp and weft should *flow* around connection points instead of being cut. Reinforcement belts also have to be welded into fabrics to compensate for loading and prevent any defects extending and ripping through the fabric.

Fibre diameters in these structures are very small and of the order of 25 microns. This makes reference and specification difficult so it is normal practice to refer to them in terms of their linear density. The main units used are:

1 tex = 1 mg/m
1 tex = 9 denier

Fire and safety requirements have to be negotiated with local building officers as there are no overall standards which currently apply at present.

Fabric technologies in building are not keeping pace with the latest developments in this area. This is partly because new products that could prove useful in the industry are very expensive, although mass use and mass production would inevitably make them far more accessible. Some polyester fabrics in sportswear are coated with a hydrophilic polymer which makes it difficult for bacteria to adhere to the surface. Although this is done for personal hygiene it does show how selective coatings can be used actively to protect a material. Another feature of the more expensive sports garments is that they can be multilayered to allow water vapour (4×10^{-4} m) to diffuse out but not allow water molecules (100×10^{-4} m) to penetrate, so being genuinely microporous and yet weatherproof. These fabrics used as roofing materials would prevent the build up of vapour which can help the inner fabric to deteriorate.

Polyester films are normally hydrophobic in their structure. The insulating fleece fabric, again polyester, is used as a *wick* to conduct moisture vapour out. The building fabric's protective shell is analogous to clothing functions, and there is scope for further improvement and development and the transfer of technologies across these fields.

Gore-Tex fabric, made by W L Gore of Delaware, uses a coating of liquid Teflon (PTFE) which bonds mechanically to fabric. It cannot adhere to any material because of its inability to bond and will remain a clean surface as debris cannot attach easily.

The latest developments in this field now investigate the use of foils which do not have a fabric base. They give high translucency and are of ethylene tetrafluorethylene (ETFE) and fluorinated ethylene-propylene (FEP) polymers. They are being used as inflated cushions or sometimes as membranes over GRP ribs for support. This is now a highly specialized area with a great deal of expertise.

See *Patterns 5* published by Buro Happold, May 1989, and the teaching pack which includes two videos; *Tension structures*, produced by the Building Centre Trust in 1990 accompanying the exhibition of the same name.

It may be useful to refer to BS 3496 *E glass fibre chopped strand mat for reinforcement of polyester and other liquid laminating systems*. Strands in the types of mat discussed are made from fibres with an average diameter of between 8 and 15 microns, with discontinuous strand length not over 25 mm. (E glass is defined as glass with not more than 1% by mass of alkali.)

10.5 Windows

Plastic frames

There is an increase in the use of uPVC windows, originating from the continental market over the last few years. Generally it appears that use of these windows is

Figure 10.1 Designers exploring and using fabric technologies will learn a great deal from the structures and fabrics used by manufacturers of tents, balloons and dirigibles: (left) changing pavilions, France; (middle) British Gas balloon, event in Bristol; (right) dirigible at RAF Cardington.

at the expense of aluminium and not wooden window frames. They are manufactured to high standards in terms of durability, thermal properties (in preventing cold bridges) and their ability to resist water penetration. Users should ensure they use manufacturers who have Agreement certificates. Attention should be paid to the use of gaskets on these windows. PVC gaskets will have plasticizers that will tend to migrate into the window frame causing blistering. Recommended gaskets use silicone and EPDM. uPVC units with 20 mm sealed glazed units can achieve U values between 2.1 and 2.6 W/m^2 K. These frames may need additional reinforcement to counter the effects of severe wind loading using the exposure charts contained in BS 6375. Stiffness can be achieved from the polymers used in the window construction coupled with the relative geometric stiffness obtained from complex extrusions which give three-dimensional stability. These window systems under accelerated ageing tests indicate a life of 25 years. There is a security problem with some of these systems where beads can be removed externally. The main dilemma in their use is a cultural one. As a product they are proved and tested, particularly for extreme weather conditions, and yet they might be thought inappropriate for the sensitive conversion of an existing building.

Plastic glazing

In some situations plastic glazing may be considered as an alternative to glass, as a security measure for example, and there is a section in BS 6262:1982 which deals with this range of materials. (See also Section 4 of the *Glazing manual* published by the Glass and Glazing Federation in May 1989 which is a useful guide.)

The three main materials used are polycarbonate (PC), polymethylmethacrylate (PMMA) commonly known as perspex, and poly(vinylchloride) (PVC). They can be used as sheet materials or profiled as hollow sections, commonly used as a roofing material.

These materials will deflect more than glass and table 2 of the *Glazing manual* gives thicknesses for these materials with respect to wind loading, area of glazing and aspect ratio. As deflection is so much greater, edge cover of at least 15 mm is recommended to prevent the pane from springing out under load and this should be increased if there is need to take precautions against vandalism. Thermal expansion is also greater than glass and will require additional clearance of 3 mm per metre of panel.

Sound insulation properties vary directly according to mass; Table 5.5 gives the mean sound reduction for single glazing and Table 5.6 gives the mean sound reduction for double glazing in BS 6262. Performance varies from a single sheet of 3 mm giving a decibel reduction of 18 dB to two sheets of 13 mm with an air gap of 250 mm giving a reduction of 49 dB.

Aspect ratios are more critical with regard to the proportional relationship of the longest to the shortest dimension and panes sized over 2 m^2 will need specialist calculation. (See the *Glazing manual*, figures 14–19, with respect to the calculation of pane thickness relative to wind speed, cover area and aspect ratio.)

See BS 6206: *Classification for safety plastics.* Manufacturers should be consulted with regard to the choice of glazing materials such as sealing compounds. The coefficient of thermal linear expansion is high and falls between 6.0 and 8.0 × 10^{-5}/K. Characteristically this gives a probable expansion of 5 mm per 1000 mm length. Temperature variations are so great that dimensional accuracy is affected when cutting at different temperatures.

This will also affect the design of rebates, and using plastics to reglaze frames designed for glass will be problematic. Rebates should also allow for glazing clearances. Generally rebates should not be less than 12 mm.

Plastics are generally more vulnerable than glass to being scratched and to suffering damage generally from aggressive environmental conditions. Cleaning methods should avoid the use of chemicals which may be harmful and should only require mild detergent with warm water, rinsing and then wiping dry.

Fine scratches can be removed by using metal polishes. Sealing compounds should always be those recommended by the manufacturer of the plastic used, and those sealants that require a primer may need an additional check with the primer manufacturer on compatibility.

Twin-walled polycarbonate sheeting is now commonly used as an alternative to glazing, especially overhead. It can be obtained as clear, opal or in solar bronze and in thicknesses from 6 to 10 mm. The twin-walled construction gives longer effective spans and useful insulation properties. Class 1 fire ratings are available from some manufacturers.

The manufacturers of the acrylic sheeting known as plexiglas have developed techniques of bending and forming acrylic up to 75 mm in thickness. Uses include the installation of completely glazed tunnels for visitors to aquaria and aircraft canopies for the British Aerospace Centre in Hamble. Sheet sizes are available up to 3800 × 1800 mm and have a light transmission of 90% or more. The slab sheets used have a gasketry system which ironically would probably work better underwater due to the pressure force component from a mass of water.

10.6 Bitumastic materials and products

Bitumen is found naturally either on its own or as the component in mineral deposits of asphalt. It can also be obtained through the fractional distillation of crude oils. Petroleum products are used to modify and liquefy bitumen for workability.

Bitumen is used in the building industry as the flexible matrix in combination with fibres, mats or aggegates to provide a range of products, and in combination with a range of materials to produce composites used for roofing felts and tiles. It can be used on its own as a sealing agent against water. Felts use glass fibres or cellulose fibres which can process the waste material obtained from timber, waste paper or rags. The felts produced from this waste should be able to absorb between one and two times their own weight in bitumen. Although regarded as a low-grade material in the building industry, which does use a great deal of waste material, it can be extremely durable especially if glass fibres or polyester are used as the reinforcing media, lasting over 20 years. It provides a cheap method of roofing. The same process is applied to the manufacture of tiles or shingles in the USA.

Other stabilizers are added to sheet products which can include talc, slate dust and silica micaceous materials, or even ash, which add to hardness, control the natural elasticity of bitumen and prevent the extreme cracking associated with low temperatures. All of these fine particles act as miniature crack stoppers to a gel that can become glass-like at very low temperatures.

It is important that the upper surface of felts exposed to the sun have integral coatings of chippings which can reflect light.

Asphalt

Asphalt is found as a natural sedimentary deposit of solid bitumen. It is a soft mineral, described as a pitchy substance composed of a mixture of hydrocarbons. It can occur as black or brown deposits and is found in England in Derbyshire, Leicestershire, and Shropshire. It is a fine-grained material which when broken exhibits conchoidal fracture. It can be used from the natural state and provides a natural paving material as in the USA, also in France where large deposits are found penetrating sandstones and dolomites. There are purer bitumens found as deposits which have no mineral distribution. The name asphalt is also commonly used (especiallly in the USA) for the artificially made asphalts which combine heavy bitumens with aggregates (which are usually gravels and sands) and are similar to the natural deposits.

Natural rock asphalt This name is given to a calcareous rock structure which is already impregnated with bitumen.

Lake asphalt This refers to naturally occuring deposits of fine mineral matter combined with bitumen.

Asphaltic cement This refers either to bitumen or refined lake asphalt which is sufficiently adhesive to form a building mastic asphalt.

Mastic asphalt This is the term used to describe graded mineral matter in combination with asphaltic cement.

The resulting solid materials which use either natural asphalt mineral deposits or bitumen, with selected particular or fibre reinforcing agents, share some common characteristics. They are elastic gels and differ from concrete (which is a rigid gel) only by the binding medium which is a bitumastic matrix. This matrix of bitumen is a thermoplastic polymer which re-softens with the application of heat, enabling it to flow to some extent around the aggregate particles. The method for deciding on ranges of aggregates or grading is much the same as concrete with detailed specifications that relate to differing performance.

Asphalt is known not only as a road surfacing material but also as a roof covering and floor finish. The difference in specifications will be marked by the size of the aggregates which, for example, are larger for roads than for footpaths or roofs. Roads and roofs do have good analogies in their detailing and the Fiat factory building in Turin was originally detailed with a testing track for cars at roof level. If the specification is good it has an extremely long life, and one advantage over other materials is that it can be repaired easily if necessary. As a fluid formed material it can be used for the covering of complex shapes and be taken up vertically to provide waterproof skirtings and tanking details.

Asphalt as a gel is derived from its matrix of bitumen which is a viscous liquid, or solid of hydrocarbons soluble in carbon disulphide, or trichloroethylene refined from petroleum and can be found naturally. It is a thermoplastic polymer. This explains its ability to soften on heating, generally between 65 and 95 °C, so it is vulnerable in direct sunlight (which can cause shrinkage and surface crazing) and in summer as ambient temperatures rise. It will be laid in a liquid form after heating to between 175 and 230 °C. It does become brittle at sub-zero temperatures. Asphalt is then a particle composite with a non-rigid matrix but, like concrete, it is more successful as a material if the grading gives a normal distribution, and a large range of different sized particles. It is ideal for the particles to be angular which gives a stiffer more interlocking structure. Unlike concrete it becomes critical if the ratio of bitumen to aggregate is too large. The particles should just touch and the voids be minimized. If there is too much bitumastic or binding medium, the asphalt will flow plastically under load and there will be far greater deformation.

Asphalts may also be modified. They can mixed with solvents or emulsified in water. In this state they are used as thinner coatings for first stage waterproofing before applying felting or as a primer to masonry before applying heavier asphaltic coatings.

Asphalt cement can be produced for repair purposes which will include reinforcing fibres such as glass. (BS 1446:1973 *Mastic asphalt (natural rock asphalt fine aggregate for roads and footways)*.)

The percentage of bitumen in the mix will be between 13 and 16% if Swiss rock asphalt is used, and 17 and 20% if other rock asphalt is used. The greatest percentage of mineral mass (between 45 and 65%) will be fine particles below 75 microns in diameter, showing the fine grain of the composite. (See Table 10.4.)

Aggregates are limited to igneous (granites, etc.) or calcareous rocks (limestones) and natural rock asphalt may be used as the fine aggregate with a bitumen content of between 6 and 10%.

It was recognized in BRE Digest 144:1972 *Asphalt and built up roofings: Durability* that flat roofs had been built essentially as composite structures in terms of the way they were layered and how each layer was bonded one to the other and the materials cannot be looked at in isolation. Building practice today advises the separation of layers which all respond at different rates thermally, and that the top surface of asphalt and other roofs should be treated in such a way as to reflect solar energy as a preventative measure. Organic fibre composites were generally recognized to be more unstable as the fibres could decay. If replaced by glass fibres, unless these were in the form of a woven mat, the structural discontinuity of fibres could cause cracking just from incorrect handling of the material while being placed.

Water vapour from the building fabric can expand in hot weather putting great pressure on asphalt which will subsequently blister.

One common detail that has changed slightly is the way in which metal flashings lap over asphalt. In warm weather the metal can attach itself to asphalt and pull the flashing away from the wall or alternatively produce a horizontal crack in the asphalt if the flashing stays in place. Tape fixed over the asphalt before the metal flashing is attached will give sufficient separation. Success in the use of these

Table 10.4 Composition by analysis of mastic asphalt

Rock type	Min./max. % by mass of mastic asphalt				
	Soluble bitumen	Retained in given BS sieve			
		0–75 μm	75–212 μm	212–600 μm	600–2360 μm
Swiss rock asphalt	13/16	45/65	5/20	5/20	0/5
All other natural rock	17/20	45/65	5/20	5/20	0/5

Source: Table 3 BS 1446 : 1973

materials is through recognition that the layered roofing system works overall as a composite material.

10.7 Polymeric components

Panels

Glass-reinforced plastics are more commonly used to provide sheet materials that have sufficient stiffness. The technology of using these materials has improved to the point where surface finishes can reproduce the reflective quality and colour of aluminium. They can generally comply with BS 476 Part 7: *Class 1 surface spread of flame*.

As an alternative, reinforced thermoplastic sheets are now available which use a glass-fibre reinforced core with rigid PVC foils on either side. A range of colours and textures is available. Although developed for shopfitting and signage applications, it is now possible to use this as an internal cladding material with a class 0 fire rating. It provides hygienic surfaces, ideal for kitchens or laboratories, due to its high resistance to chemicals.

Although there is a great range of plastics components in the building industry, the choice and specification of the polymers used will already have been determined by manufacturers for particular products. Consequently, there is less need for a detailed knowledge by professionals who will use British or European standards to determine the quality of those components. Extra information on flexible sheets, flooring resins, coatings and other polymers can be found in *MBS*: *Finishes* which also contains a section on adhesion.

10.8 Health

Chemical products regarded as hazardous in the Health and Safety Commission's *The control of substances hazardous to health in the construction industry* include:

Solvents Toluene, xylene, trichloroethane. Used in paints, laquers, glues, strippers as thinners. They may also be present in tile fixing systems and resin systems used on site. They can be harmful when inhaled and can cause dermatitis. Exposure should be minimized with the avoidance of spray application and with good ventilation. Impervious protective clothing is recommended together with the use of barrier creams. (Health Hazard sheet 5, HSE leaflet IND (G) 72L.)

Resin systems Isocyanites, e.g. toluene di-isocyanate, diphenylmethan di-isocyanate, polyurethane coatings and adhesives. These are respiratory irritants and can initiate sensitivy and asthma. Epoxy applications are severe irritants and sensitizers and are also toxic, affecting the liver. Their use is mainly in adhesion. Polyester can produce a styrene vapour which is toxic and can be inhaled causing liver damage. It is also regarded as a narcotic and can irritate eyes. Spray application of resins is discouraged in favour of brush or roller application. If spray applications have to be used then breathing apparatus should be worn with impervious coverall. MDI is also present in roof-sprayed thermal insulation. Natural ventilation should be supplemented with mechanical air extraction.

Washing and eating facilities should be situated away from site operations. (Refer to GN 16 *Isocyanates* and GS5 *Entry into confined spaces.*)

Reference

1 H L Malhotra *The behaviour of polymers used in building construction* BRE Fire Research Note 1071.

V COMPOSITES

11 Introduction

11.1 Anatomy and microstructure of wood

In their book *Engineering Materials 2*, Ashby and Jones point out that timber is analogous to a *foamed fibrous composite*, chiefly because of the material's ability to absorb energy and behave as a tough material. Foams are composites of solids and gases.

Architects and designers generally regard timber as a graded commodity to be used, and the form in which it arrives, as neat, sawn rectangular sections with a minimum amount of knots and defects which disguise its real origin as a tree or *woody plant*, which is a better botanical description. Of course timber does not always have to be used as sawn sections. Much vernacular architecture and building in this country even well into the eighteenth century used round pole sections for trusses and sometimes joists, and this is still commonplace practice throughout the world. This is a sensible use of timber as cell walls are not cut through and exposed to moisture and this must help in durability. De-barking is advisable to harden the outer living layer and remove the zone which is a natural home for many pests. Exposed trusses and other details may not necessarily need to be sawn sections and it is an option worth considering. However, architects and engineers are not used to detailing whole wood sections, and have been known when preparing drawings (especially when drawing details that show the use of thinnings in pole construction), to show a consistent section throughout the length of the member. This ignores the fact that timber will taper towards the top of the tree crown.

Trees are exogens, that is they display outward growth in the living outer cambium layer just below the surface of the bark. New cell growth occurs in spring and summer each season between the existing woody stem and the bark. As spring growth is large celled and summer growth small celled, this gives a patterned differentiation commonly known as the *annual rings*. The age of timber can be roughly estimated by allowing for a growth in girth of 25 mm per annum. Young trees grow faster initially and if their early growth is restricted by other trees, age can be underestimated.

Figure 11.1 The de-barking of poles at the Parnham Trust.

The heartwood is the structural part of the tree and provides a rigid support for all the branches. It is usually darker than the surrounding sapwood which is *live* in terms of carrying moisture through the tree. This outer sap zone can be up to about 50 mm in width. The heartwood is generally more durable as it contains compounds that act as antifungicidal agents, which is especially true of western red cedar. If seasoned in the open air over a long period of time, these compounds may wash out of the heartwood and then decay will occur.

The term *grain* is used in describing timber which reflects

Figure 11.2 The seasoning of boles of timber which have been sawn 'through and through' in the Pyrenees. The careful stacking under load ensures the timbers season with the minimum of distortion.

Figure 11.3 The large planks have been sawn vertically to produce 'quarter sawn' square sections which are carefully stacked. At the top right hand are planks with 'wane', edges showing the natural profile of the tree trunk.

the directional alignment of the axial elements of a tree. If the grain is parallel to a log or plank, the element will be known as *straight grained*. This consistent patterning will be altered by local features, branches, etc. which in their growth will give a directional change to the grain. The expression *figure* is often used when describing local variations of colour and structure which might be used to describe the differences between the heartwood and softwood. Wood will also change its appearance according to the direction of cut. There are three main sections to recognize, *tangential, radial* and *veneers* which are usually a rotary cut around the tree bole. *Quarter sawn* timber has a more lustrous appearance than *tangentially sawn* timber.

Burrs

Burrs are used as a descriptive term for walnut but result from the swellings of trees on the lower part of the bole; the resulting pattern, although a deformation in terms of the tree, is regarded as being highly decorative.

Sawing

In sawing timber care should be taken to remove dust locally from the work area and operatives given masks. Dust from timber can be a major irritant, especially to the mucous membranes, which could cause nosebleeding, asthma and dermatitis. Hardwoods such as abura, agba, E I satinwood, iroko, makore, mansonia, obeche, opepe, peroba and teak can be especially hazardous. (*Timber*, H. E. Desch, revised by J. M. Dinwoodie, 1981.)

Microstructure

Sections across the tree show the holo-cellulose (see later) structures (65–80% of the wood) that convey liquid. On closer examination these tubes have a geometry which is roughly hexagonal and give a stiffening effect to the tube. These tubes are regarded as composite structures by being bound together with the natural polymer lignin (20–35% of the timber). There are various other minerals present as well as organic chemicals in solution. *Holo-cellulose* is a term used to describe cellulose and other carbohydrates or hemi-celluloses. Cellulose is basically a linear polymer of individual glucose molecules. The lignin content is of amorphous three-dimensional polymers. So cellulose fibres are held within a matrix of lignin, hemi-cellulose, water and other compounds.

The structure of the tracheid or cell wall or fibre is many-layered and can be regarded as a laminated composite structure, with the layers wrapping around each other in counter diagonal configurations. The cell wall is actually composed of four layers with a primary wall and then a three ply overlaying structure. The diagonal growth of the cell walls is usually described as helical in arrangement and with the help of the electron microscope, the structure of the helical arrangement can be seen as giving a distinct internal stiffening *ribbing* to the main walls. The tubular cellulose structures are strengthened by this ribbing but still strongly orientated and have directional strength, although held in the matrix of lignin. This structure has been described as parallel to the macro features of steel reinforcement in concrete, i.e. steel in a matrix of cement and aggregates, although the timber structure is a more complex system of composites, involving fibres and laminations in a matrix. With such a variety of structural configurations, it is no surprise that timber can respond to a variety of applied stresses, in different directions.

Water or moisture content

The water content of trees should be looked at in two ways. First, cell walls are conduits for water to reach the growing areas of trees and will suffer water loss immediately a tree is felled. Even after the source of water is removed water will still remain in the cell wall cavities, amounting to 25–35% of its oven dry weight. At this point timber is stable dimensionally and this level of water content is known as the *fibre saturation point*. It is when further water is lost that shrinkage of the timber starts which can eventually lead to sectional distortion. It is desirable in buildings to reduce the moisture content down to below 20% which is the dry rot safety line and so shrinkage of sections then becomes inevitable.

Conversely, timber has the ability to take up moisture if there is theoretical room in the cell walls, especially if the atmosphere is moisture saturated, the dynamics of the equalization of water vapour pressure will force moisture into the structure of timber to achieve equilibrium with external conditions. The problem is that once a building is constructed and the timbers specified for those conditions, that building must maintain a stable relative humidity to prevent the re-saturation of the material. Inevitably this is impossible to achieve under most conditions and there is seasonal variation for timber in buildings which means the material will swell and contract as conditions alter.

The moisture content limit for wood used for a variety of purposes is given in Table 11.1.

Shrinkage in timber in the green state varies from 0.1 to 0.2% longitudinally and from 2 to 10% across its width. Shrinkage will also vary according to how timber is sawn, and shrinkage can double if timber is sawn tangentially rather than radially. Shrinkage increases the further from the centre the cut is made; the higher moisture content will be on the outer growth side of the tree and will then suffer

Figure 11.4 Building in the grounds of Leeds Castle, Kent: (left) bundles of timber remaindered from sawing through and through and quarter sawing; (right) some sections can be used directly for cladding. By using log sections, this prevents the exposure of cut grain and provides a more durable cladding.

Table 11.1 Moisture content limit of wood by intended use

Wood usage	Moisture content (%)
Carcassing	22
Flooring/utility soft woods	18
Decorative hard woods	10–12
Underfloor heating	8–10
Continuously heated	10 ± 2
External joinery	17 ± 2

Source: TRADA Leaflet W1 4/12 1979

the greater water loss. This is noticeable on large board sections if cut tangentially, causing distinct cupping.

Hardwoods are more stable in shrinkage, partly due to the effect of medullary rays which are stabilizing fibres running in the opposite direction to cell wall structures, and partly because of greater cell wall thicknesses leading to increasing density of the material.

Kiln drying In order to achieve the correct moisture content timber is seasoned naturally through air drying or artificially by kiln drying.

Kiln drying also alters the colour of timber. Generally the moist heat darkens timbers and these effects are sometimes counteracted by adding lime to lighten the colour, fuming with ammonia to increase darkening and remove pink and red shades, or straightforward bleaching with hydrogen peroxide.[1]

Reference

1 H E Desch revised by J M Dinwoodie 1981 *Timber*.

12 Timber

12.1 Classification of hardwoods and softwoods

The chief difference between hardwoods and softwoods is in their botanical classification. They are different species, reflected in their different methods of propagation and in their microstructure. Hardwoods are Angiosperms with their seeds protected in fruits and are broadleaved. They are also often deciduous and lose their leaves annually. Softwoods are Gymnosperms with naked seeds, recognizable as cones, and their leaves are needle-like and often coniferous or evergreen. However, these are very general descriptions and some softwoods, European larch for example, will shed all of their leaves in one season instead of a gradual replacement. Some hardwoods are evergreen, holly being one common example.

Common names which are botanically incorrect are used for many timbers and, to avoid confusion, the Latin names of timber should be used. These are clear, giving the generic name first, e.g. *Pinus* which always has a capital letter and will indicate the particular species. There may be a last letter which denotes the name of the botanist who categorized the species. In the case of *Pinus sylvestris* the name will indicate that the timber belongs to the pine family. Specifications should always be exact, especially when timber of a particular structural grade is wanted. A common assumption made by architects is that hardwoods are all generally more durable than softwoods. This is not the case, and one architect related to me how after an investigation into why hardwood sills had rotted through after only a few years (after being specified as hardwood on the drawings and in the specification) the Building Research Laboratory confirmed that a non-durable hardwood had been chosen by the contractor, and that simply to state *hardwood* as a description was not adequate and that an actual species should have been specified. A common detail found in London streets as a surfacing material is the use of pitch pine blocks as an alternative to cobbles, often found in good condition after over 100 years of external use. This shows how durability and species is not necessarily related to the hardwood/softwood classification.

It is normal for timber merchants to sell timber as unclassified softwood and, if not stress graded, to have classifications of *carcassing* or *structural* or *best* for joinery. Often the only softwood actually named is *parana pine* which is not even of the *pinus genera* family and is really *Araucaria angustifolia* from South America. The only native pine in Britain is *Pinus sylvestris* which is commonly known as *redwood*.

Deal is another confusing term given to timber and is now often used when referring to timber of a lower quality than pine, commonly fir. The origin of the name is in the sizing of timber when deal meant a section of timber approximately 9 × 3 inches and at least 6 foot long.

Common name	Latin name
Red deal Scots pine	*Pinus sylvestris*
Yellow deal Yellow pine	*Pinus mitas* or *strobus*
White deal Norway spruce	*Abies excelsa*

In addition, parameters for choice may be influenced by whether the timber is a hardwood and if it has been taken from a country or region which has an active management policy with regard to the provision of sustainable resources. Wherever possible alternative hardwoods from temperate climates or even softwoods should be considered first. *The good wood guide* is a publication from the Friends of the Earth which gives the background for responsible choice and lists suppliers and other organizations which have a defined policy for choice.

There is general ignorance on timber species. Architects and designers should be able to recognize and specify those woods given in Table 12.1.

Table 12.1 Common softwoods and hardwoods

Softwoods		Hardwoods	
Scots pine	*Pinus sylvestris*	Oak	*Quercus robur*
Norway spruce	*Abies excelsa*	Beech	*Fagus sylvatica*
European spruce	*Picea abies*	Ash	*Fagus excelsior*
Yellow pine	*Pinus strobus*	Mahogany (American)	*Swietenia macropylla*
Parana pine	*Araucaria angustifolia*	Mahogany (African)	*Khaya ivorensis*
Pitch pine	*Pinus palustris*	Teak	*Tectona grandis*
Douglas fir	*Pseudotsuga menziesii*	Iroko	*Chlorophora excelsa*
Western red cedar	*Thuja plicata*		

The identification of timber

Timber can only be properly identified by looking at its anatomical structure. A hand lens is sufficient for hardwoods but a microscope is needed to identify softwoods. Other methods of detection use density differences, hardness, colour and smell.

As a property, density is often the best parameter to describe timber in terms of its strength, behaviour and likely durability.

Structural differences in hardwoods and softwoods

Softwoods generally have a larger tracheid or cell structure and the structural emphasis is vertical although the tracheids are not continuous and represent the mature thick lignified cell wall which is emptied of its protoplast and has become a non-living cell wall. Although in some softwoods they contain resin, either as deposits on cell walls or bridging across cell walls, they have a straightforward anatomy. As the cell walls are thin, they are easier to cut than those of hardwood.

Hardwoods have many other features in their anatomy and are more complex structures having five cell wall types as well as *rays*, so looking at the grain there appears to be a greater diversity of structure. In addition, hardwoods have fibre cells in the cell walls and as these increase so does the density and strength of the timber. Lignum vitae (used for making the spherical *woods* in the game of bowls) is extraordinarily dense as the cell cavities are filled by fibre growth in the cell walls and is one wood that will sink in water.

Hardwoods are also distinctive because of their medullary rays which have a clear structure that runs across the grain. It is not necessarily visible as a feature in all hardwoods but is very distinctive in oaks. Depending on how oak is cut this feature will either appear as a dark regular fleck, parallel to the grain on tangentially cut surfaces or as a light, more irregularly shaped, fleck on surfaces that have been cut radially.

This differing structure in hardwoods may sometimes affect how the timber will wear, and coarser more visible structures will be more differentiated in this respect. There will also be uneven wear between the annual rings as the softer material associated with spring growth wears first, leaving the harder, more closely aligned, summer growth rings.

12.2 Structural classification of timbers and commercial grading

Strength of timber

Timber has a natural directional strength due to its cellulose tube structure, and crack propagation will take place preferentially along the grain.

Values for the strength of timber have often been determined from the testing of small-scale specimens and then a safety factor may be incorporated. (Factors of safety are determined from an agreed probability of failure which is taken as 1:40 in CP 112 Part 2:1971.) Safety factors for timber will generally work out at 2.25 for individual components tested and can reduce to 2.00 if five or more components are put through a testing procedure. Strength in compression and tension generally increase with density and so the hardwoods, being denser than softwoods, are invariably stronger. The mechanical properties are measured in terms of establishing Young's modulus, the strength, and finally the fracture toughness of the material. There are great variations in these figures as the material is obviously not consistent and has many flaws. Fracture toughness is becoming more important as an indication of the strength of timber as it measures the ability of the material to propagate a crack under stress from an initial defect.

As timber is not a homogenous material it becomes important to quantify the number and nature of the defects and to predict the likelihood of fracture under loading. The greater the length of timber, or the larger the section size, the less efficient it will be. Consequently, efficient timber design is optimized if lightweight structures made up of smaller elements are used. Laminated beams, although they appear to be substantial structures, will have additional likely defects in between their laminations from voids, adhesion failure or stress concentrators in the form of fixing details, and profiling of the section.[1]

In the future the development of better testing procedures and grading of timber, coupled with a larger understanding of lefm (linear elastic fracture mechanics) may be used more to determine the strength of timber.

Commercial grading

Defects in timber include longitudinal separation of fibres which may appear on faces edges or ends of softwood and also checks, shakes or splits. The *knot area ratio* (KAR) is critical to strength and is the ratio of the projected cross-sectional area of the knot to the face of the timber. It is not the visible cross-sectional area on the timber, but the elevational cross-section, which would of course be far larger. Example gradings and properties are given in Table 12.2, and Tables 12.3 and 12.4 give more detailed information.

There are also categories of distortion to whole sections of timber which include bow, spring, twist and cup as well as resin pockets, bark pockets and insect damage.

Figure 12.1 This section across the bole shows the radial shakes which would show as splits on a cut section of timber. The centre of the tree shows cup shakes. This tree fell in the great storm of 1989 and most of the timber would be unusable.

It is important to know that when timber is delivered, it is still acceptable to have a sample equal to or less than 10% which can exceed the permissible limits given, of which 3% can exceed the limits by one third. Then 'the parcel shall be considered as complying with the grade'. The only exception is if the parcel is less than 50 pieces

Table 12.2 Commercial grading of wood

Criteria	Grade	
	General structural (GS) European S6	Special structural (SS) European S8
Knots	TKAR 1/3	TKAR 1/5
Slope of grain	1 : 6	1 : 10
Width of growth rings (mm)	10	6
Width of fissures (mm)	10	10
Wane	Up to ⅓ of any side	Up to ¼ of any side

Figure 12.2 Section of timber from a sculpture by David Mach entitled 'Teddy Bear's picnic' and showing classic star-shaped 'heart shakes' in the centre of the timber.

Table 12.3 Permissible limits for visual stress grading

Characteristic		Grade: GS	Grade: SS
Knots	*Rectangular sections*	If MKAR is less than or equal to $\frac{1}{2}$, then TKAR shall be less than or equal to $\frac{1}{2}$ *or* If MKAR is greater than $\frac{1}{2}$, then TKAR shall be less than or equal to $\frac{1}{3}$	If MKAR is less than or equal to $\frac{1}{3}$, then TKAR shall be less than or equal to $\frac{1}{3}$ *or* If MKAR is greater than $\frac{1}{3}$, then TKAR shall be less than or equal to $\frac{1}{5}$
	Square sections	TKAR shall be less than or equal to $\frac{1}{3}$	TKAR shall be less than or equal to $\frac{1}{5}$
Slope of grain		1 in 6	1 in 10
Rate of growth		Average width of annual rings not greater than 10 mm	Average width of annual rings not greater than 6 mm
Fissures	*Not through the thickness*	Fissures less than 10 mm in depth may be ignored	As GS
		Not greater than $\frac{1}{4}$ of the length of the piece or 900 mm, whichever is the lesser	Not greater than $\frac{1}{4}$ of the length of the piece or 600 mm, whichever is the lesser
	Through the thickness	On any running metre not more than one fissure of the maximum length	As GS
		Not greater than 600 mm. If at the ends with a length not greater than $1\frac{1}{2}$ times the width of the piece	Only permitted at the ends with a length not greater than the width of the piece
Wane		$\frac{1}{3}$ of the thickness, $\frac{1}{3}$ of the width for full length; $\frac{1}{2}$ in any 300 mm length	$\frac{1}{4}$ of the thickness, $\frac{1}{4}$ of the width for full length; $\frac{1}{3}$ in any 300 mm length
Distortion	*Bow*	In a thickness of 35 mm not greater than 30 mm. In a thickness of 44 mm not greater than 25 mm. In a thickness of 75 mm not greater than 10 mm. (Intermediate sizes may be determined by interpolation)	As GS
	Spring	In a width of 60 mm not greater than 10 mm. In widths of 250 mm not greater than 5 mm. (Intermediate sizes may be determined by interpolation)	As GS
	Twist	Not greater than 1 mm per 25 mm of width	As GS
	Cup	Not greater than 1 mm per 25 mm of width with a maximum of 6 mm	As GS
Resin pockets and bark pockets	*Not through the thickness*	Unlimited, if shorter than the width of the piece. Otherwise the same limits as for size of fissures	As GS
	Through the thickness	Unlimited, if shorter than half the width of the piece. Otherwise the same limits as for size of fissures	As GS
Insect damage		Worm holes and pin holes are permitted provided they comply with the requirements of clause 5 of BS 4978. No active infestation is permitted. Wood wasp holes are not permitted	As GS

Note: Distortion will be influenced by the moisture content of the softwood at the time it is measured. Where for a particular reason differing limits from those indicated are required, these should be subject to special arrangements between supplier and purchaser.

Source: Table 1 BS 4978 : 1988

Table 12.4 Visual requirements for machine grades

Characteristic		Grade		
		MGS, SC1, SC2, SC3	MSS, M50, SC4	M75, SC5
Fissures	*Not through the thickness*	Fissures less than 10 mm in depth may be ignored	As MGS, etc.	As MGS, etc.
		Not greater than $\frac{1}{4}$ of the length or 900 mm, whichever is the lesser	Not greater than $\frac{1}{4}$ of the length or 600 mm, whichever is the lesser	Not greater than $\frac{1}{5}$ of the length or 500 mm, whichever is the lesser
	Through the thickness	On any running metre not more than one fissure of the maximum length	As MGS, etc.	As MGS, etc.
		Not greater than 600 mm, if at the ends with a length not greater than $1\frac{1}{2}$ times the width of the piece	Only permitted at the ends with a length not greater than the width of the piece	Only permitted at the ends with a length not greater than the width of the piece
Wane		$\frac{1}{3}$ of the thickness, $\frac{1}{3}$ of the width for full length; $\frac{1}{2}$ in any 300 mm length	$\frac{1}{4}$ of the thickness, $\frac{1}{4}$ of the width for full length; $\frac{1}{3}$ in any 300 mm length	$\frac{1}{8}$ of the thickness, $\frac{1}{8}$ of the width for full length
Distortion	*Bow*	In a thickness of 35 mm not greater than 30 mm. In a thickness of 44 mm not greater than 25 mm. In a thickness of 75 mm not greater than 10 mm. (Intermediate sizes may be determined by interpolation)	As MGS, etc.	As MGS, etc.
	Spring	In a width of 60 mm not greater than 10 mm. In widths of 250 mm not greater than 5 mm. (Intermediate sizes may be determined by interpolation)	As MGS, etc.	As MGS, etc.
	Twist	Not greater than 1 mm per 25 mm of width	As MGS, etc.	As MGS, etc.
	Cup	Not greater than 1 mm per 25 mm of width with a maximum of 6 mm	As MGS, etc.	As MGS, etc.
Resin pockets and bark pockets	*Not through the thickness*	Unlimited, if shorter than the width of the piece. Otherwise the same limits as for size of fissures	As MGS, etc.	As MGS, etc.
	Through the thickness	Unlimited, if shorter than half the width of the piece. Otherwise the same limits as for size of fissures	As MGS, etc.	As MGS, etc.
Insect damage		Worm holes and pin holes are permitted provided they comply with the requirements of clause 5 of BS 4978. No active infestation is permitted. Wood wasp holes are not permitted	As MGS, etc.	As MGS, etc.

Note: Distortion will be influenced by the moisture content of the softwood at the time it is measured. Where for a particular reason differing limits than those indicated are required, these should be subject to special arrangements between supplier and purchaser.

Source: Table 2 BS 4978 : 1988

of timber. This shows the care that must be taken in sorting timber on arrival on site and not expecting it all to comply with the grade. This margin of waste should be allowed for in a bill of quantities.

The slope of the grain is important with regard to establishing the actual strength of the timber. Timber has directional strength, due to the nature of the bundled cellulose tubes and any deviation from being absolutely straight will affect its properties. A scribe tool is used to determine this. It will run freely down the grain, which is also likely to be at a different angle to the visual slope — hence the need for a tool.

Wane refers to an edge of timber missing, usually where the timber section forms a boundary with the original tree bole bark. Wane often coincides with a split along the grain.

Laminating grades There are three grades: A as top, B and then C:

- LA is more stringent than SS.
- LB is similar to SS but less rigorous for TKAR.
- LC is slightly better than GS for the slope of grain qualities but less rigorous for TKAR control.

There are other conditions for laminated timber where directional strength and hence slope of grain is important, and defects such as wane are not tolerated as they will give an immediate weakening in a composite. Resin pockets are not allowed as they affect adhesion.

Much timber is now machine graded and BS 4978 sets out the operable limits for mechanical handling.

Figure 12.3 This rejected stack of timber in the Pyrenees shows the characteristic 'cupping' of unsupported 'through and through' sawn sections.

Machine grades Four examples of machine grades are:

MGS MSS M50 M75

The numbers give an idea of the reduction of strength relative to the amount of timber that is defect free, so M75 = 75% free of defects.

Other machine grades are:

Table 12.5 Grade stresses and moduli of elasticity

Strength class	Bending parallel to grain (N/mm²)	Tension parallel to grain (N/mm²)	Compression parallel to grain (N/mm²)	Compression perpendicular to grain* (N/mm²)		Shear parallel to grain (N/mm²)	Modulus of elasticity (N/mm²)		Approximate density† (kg/m³)
							Mean	Minimum	
SC1	2.8	2.2‡	3.5	2.1	1.2	0.46	6800	4500	540
SC2	4.1	2.5‡	5.3	2.1	1.6	0.66	8000	5000	540
SC3	5.3	3.2‡	6.8	2.2	1.7	0.67	8800	5800	540
SC4	7.5	4.5‡	7.9	2.4	1.9	0.71	9900	6600	590
SC5	10.0	6.0‡	8.7	2.8	2.4	1.00	10 700	7100	590/760
SC6§	12.5	7.5	12.5	3.8	2.8	1.50	14 100	11 800	840
SC7§	15.0	9.0	14.5	4.4	3.3	1.75	16 200	13 600	960
SC8§	17.5	10.5	16.5	5.2	3.9	2.00	18 700	15 600	1080
SC9§	20.5	12.3	19.5	6.1	4.6	2.25	21 600	18 000	1200

*When the specification specifically prohibits wane at bearing areas, the higher values of compression perpendicular to the grain stress may be used, otherwise the lower values apply.
†Since many species may contribute to any of the strength classes, the values of density given in this table may be considered only crude approximations. When a more accurate value is required, it may be necessary to identify individual species. The higher value for SC5 is more appropriate for hardwoods.
‡Note the light framing, stud, structural light framing no. 3 and joist and plank no. 3 grades should not be used in tension.
§Classes SC6, SC7, SC8 and SC9 will usually comprise the denser hardwoods.

Source: Table 9 BS 5268 Part 2: 1988

SC1 2.8 N/mm^2
SC2 4.1 N/mm^2
SC3 5.3 N/mm^2
SC4 7.5 N/mm^2
SC5 10.0 N/mm^2
SC6 12.5 N/mm^2
SC7 15.0 N/mm^2
SC8 17.5 N/mm^2
SC9 20.5 N/mm^2

Whatever the grading system every piece of timber should be marked with the correct grade and this is a requirement of BS 4978.

Table 12.5 demonstrates that the grades of stresses increase with the density of timber. Density is often an indicator of strength.

Structural use of timber

Armed with the information with regard to the quality of timber, BS 5268 *Structural use of timber* the Part 2:1989 *Code of practice for permissible stress design, materials and workmanship* covers timber/ply/glued/laminated timber and tempered hardboard. There are also recommendations for the design of nailed/screwed/ bolted/connected/glued joints. Timber cannot be thought of in isolation as a material particularly as the method of jointing, restraint and fixing details are integral to satisfactory timber design. The choice of sections is also determined by the usage. For example, tongued and grooved boarding will give good lateral stability and sideways distribution of load, useful for designing against racking in wind loading and uniform loading generally. Butt fixed boarding will have to be designed to take full loading conditions.

In the structural design of timber, the bearing stresses that can be taken vary with the end conditions, and as the bearing is reduced so will be the capacity of the timber. The less the bearing, the greater the modification factor. Ideally, bearings should be about 150 mm. It is not helpful when minimum bearings called for in the Building Regulations (1985 Part A *Structure*) are about 100 mm as this relates better to steel and concrete. The significance of this is the ability to keep the structural timber section in alignment. Table 12.6 shows that the modification factor stays at 1.0 for bearings of 150 mm or more but then increases to 1.74 for a 10 mm bearing, almost doubling the bearing stress. The effective span is between the centres of the bearings. The ideal depth to breadth ratio can be affected and can vary from two (no lateral support) to seven (held in line for example by continuous sheathing or joist connections).

Holes should not be made in timber. They act as stress concentrators and notches/drillings all reduce strength. If

Figure 12.4 'The Collapse of Time' by John Hejduk. Without the careful grading of timber, it is not possible to achieve the building of structures that are true and easily worked, as shown in this sculpture, loosely based on American box car construction and built under the direction of Gill Djiai in the Architectural Association workshop.

Table 12.6 Modification factor K_4 for bearing stress

Length of bearing (mm)	Value of K_4
10	1.74
15	1.67
25	1.53
40	1.33
50	1.20
75	1.14
100	1.10
150 or more	1.00

Source: Table 18 BS 526 Part 2: 1988

unavoidable they should be equal to or less than 25% of the width of the member and on the neutral axis between 25 and 40% of the length from the support or end of the timber member. (See *MBS*: *Structure and Fabric Part 1*.)

Section 6.1 of the BS 5268 Code of Practice gives theoretical advice on the design of timber frame systems but more immediate help is given in the Timber Research and Development Association (TRADA) publication *Timber frame construction* 1988.

Structural failure

The strength of timber longitudinally is generally 40 times that across the grain. Failures are a repercussion of crumpling or buckling of the cell walls, which in compression failures can be seen as a deformation angle of about 30 to 40 degrees generating a total slip. These buckling failures are a way of the timber taking stress and will not necessarily be the cause of immediate failure. However, this permanent deformation of the material may cause failure under lesser loading conditions later. This is an example of overstressing having an immediate effect which is recognizable visually.

There are situations where loads can be applied to timber, deliberately not reaching its ultimate strength, and after the load is released, further applied loads can be taken by the timber beyond its predicted safe stress. This is a form of pre-stressing and acts by the work energy of the original loading causing structural changes that can physically store some of this energy. Under further loading this stored energy can be released and act in an opposite manner encouraging resistance.

12.3 Timber decay

Deterioration occurs in timber as part of a natural process in returning dead timber to the forest floor and to recycle essential nutrients. To assist this process it is a good source of food chiefly for beetles and fungi, which have short-term interest in timber in perpetuating their own existence, although the whole process is part of a much longer cycle of events. It is this natural process which is often lost sight of in building as the timber becomes a product, and not easily identified as a natural material still capable of taking part in this degenerative process.

There are then two major forms of decay: first through fungal, and second through insect attack. The BRE report *Recognising woodrot and insect damage in buildings* (1987) is a long awaited and easily referenced small handbook which aids identification. It also lists other common insects and mould which are not damaging to the fabric of the building but are a nuisance and still indicators of localized damp or condensation which need preventative measures. The handbook has full colour pictures and makes the job of identifying wood rot and insect damage much easier.

This guide also lists the equipment needed to take on site to make a proper evaluation. These should include a knife or bradawl to probe timbers, a hand lens (many distinctive features in the guide are shown at 10 times life size) a moisture meter, torch, bags for specimens, crowbars for lifting floor boards and hammer and nails to replace them. Fuller inspections will need ladders and protective clothing.

External surveys are critical in evaluating defects which may allow the ingress of water. These will include the condition of the roof, noting any missing tiles or slates, blocked or leaking gutters and downpipes, failure in flat roofs, deteriorating brickworks or renders, condition of damp proof courses, blocked air bricks, and running overflows which may be saturating the building fabric.

Fungal attack

Fungal attack divides into wet rot and dry rot. (See TRADA leaflet TBL37, 1986.)

The two main types of fungal attack in buildings that seriously weaken timber are dry rot and wet rot. The life cycles of fungi all start from a fruiting body which produces great numbers of spores. The dry rot fungus can produce spores that can approximate between 2 and 5 million per square foot for a period of several days and which looks like a fine orange dust. Spores then have be deposited on to moist surfaces to germinate where they produce fine threads or hyphae. These interlock into sheet-like growths known as mycelium and when the mycelium is mature it will produce new fruiting bodies.

Dry rot Dry rot, *Serpula Lacrymans*, is the most damaging fungus and spores thrive on softwood timber with a moisture content of between 20 and 25% in temperatures between 7 and 27 °C. The hyphae can grow and extend

Figure 12.5 Timber decay is a natural process and a host of beetles and fungi are equipped to return it to the forest floor. Our use of timber must then use 'unnatural' means for preservation, a rigorous control of moisture content and methods of preservation which make the timber unpalatable or toxic to these small life forms, and potentially dangerous to human respiratory systems.

through plaster and brickwork and so may travel some distance from the original source of the moisture. From the artificial infection of timber observed by the Forest Products researchers, tests showed a spread of 350 mm in 4 months, 1075 mm in 9 months and 2150 mm in 17 months. This shows that dry rot should not be ignored and should expect immediate attention once found.

The fungus is able to manufacture water during the breakdown of timber and so can survive at distances that seem remote from the initial sources of damp within a building that may have given rise to the attack. There is a characteristic musty smell to buildings affected with dry rot, and affected timber tends to bulge and crack. The fungus digests the lignin which binds the cellulose fibres of timber and leaves timber in a state with characteristic cubical cracking, which can be crumbled easily in the fingers. The hyphae can be white or grey when young and from fine threads can develop into plant like growths as thick as a pencil. The mycelium sheets are white but can be discoloured to yellow or grey if exposed to light. The fruiting bodies are fleshy and a yellow-brown colour, becoming rusty red when in the middle of spore production.

The complete infestation has to be uncovered and all affected wood cut back to 500 mm past the last identified piece of affected timber. The sources of moisture which caused the rise in moisture content of the timber has to be located and proper measures taken to repair downpipes, gutters or leaking pipes that might have contributed to water penetration. Affected timber must be bagged and taken off site. The existing structure should be treated with preservatives to prevent any further outbreak. Walls should only be irrigated with solution if surface treatments are reasoned not to be fully effective or to provide a distinct barrier against valuable structures which may need future protection. It is common in London to have wall structures which are of timber frame construction with brick infill. These are particularly vulnerable to the spread of dry rot and this kind of radical protection could then be justified. The fungus cannot obtain any nourishment from dry walls, so as long as the source of supply of moisture has been cut off this kind of treatment should not be necessary. The saturation walls could lead to other problems, including fresh efflorescence and it will take a long time to dry out.

The solutions are usually water based and contain

fungicides. Preservatives usually contain 'Biocides' which act as fungicides as well as insecticides. All new replacement timbers should also have been treated.

The fungus is killed if the temperature locally can be brought up to 40 °C for at least 15 minutes. It used to be common practice to use heat from a blow lamp but because of the risk of fire and the relative effectiveness, this practice was discouraged. In view of the concern about preservatives, guidelines for blow lamp heating are reproduced here. The right critical temperature will be reached by heating a wall:

for 10 minutes	to affect a depth of 65 mm
for 1 hour	to affect a depth of 112.5 mm
for 2.5 hours	to affect a depth of 175 mm

Wet rot Wet rots are not so structurally damaging as dry rots and usually have a skin of relatively undamaged timber. They need a higher moisture content in timber to survive, above 25%. The most common is *coniophora puteana* or the *cellar fungus*. The strands are dark, usually brown or black, and the mycelium is yellowish brown. The fruiting bodies are not so common as dry rot and are olive brown in colour with a cream margin. Again the extent of the growth has to be determined and all affected timber stripped out. The sources of damp have to be dealt with first before replacement of any adjacent existing timbers with timber treated with preservative.

Figure 12.6 The School of Architecture, Lyons by Jourda and Perraudin. The carefully cast junctions which house the laminated timber sections and minimize the transfer of stress are reminiscent of nineteenth century technologies.

Beetle attack

Beetles have a life cycle which starts from eggs which develop in days to larvae which start to eat their way through timber. As the larvae get fatter and healthier, the diameter of the borehole increases. They digest the timber they eat leaving behind *frass* which is different for every beetle type and a good way to identify beetles. This process until the larvae pupate can take five years. The pupae will take about six weeks to hatch and after only two or three weeks will be able to lay eggs. Active infestation can usually be seen by the amount of bore dust present. There are a large number of beetles which produce holes varying in size from 2 mm to 6 mm, although holes made by stag beetles can be 20 mm in diameter. Boring tunnels usually follow the direction of the grain, and so substantial damage is hidden. Holes can also be made by moths, sawflys, bees and wasps. Once evidence of beetle attack is discovered, a complete survey of the building must establish the extent and degree of structural damage. Timber which has any

Figure 12.7 Beetle attack at the base of a column in the Kiyomizu-dera Temple, Kyoto. Damage results directly from structural weakening of timber as density is reduced.

evidence of infestation should be removed and adjacent timbers should be sterilized. The exact chemicals used will vary and depend on the type of beetle, and these are carried in the medium of organic solvents, emulsions or pastes. The most common sources of infestation are as follows:

Common furniture beetle This beetle affects softwoods and European hardwoods. It is common in furniture. The holes are circular and between 1 and 2 mm in diameter. It can easily be misidentified as deathwatch beetle, powder post beetle and even moths. Four other species produce similar holes. It affects timbers which may be damp and are adjacent to external walls or around loft hatches. It will only attack old plywoods which have animal-based adhesives. The frass which is digested bore dust is gritty when rubbed between the fingers.

Powderpost beetle This infests the sapwood of European and tropical hardwoods and is likely to be present when the timber is first delivered as infestation is usually passed on in storage rooms. The holes are circular and between 1 and 2 mm in diameter. The bore dust is very soft and talc-like.

House longhorn beetle This beetle is very localized in its attacks, mainly in Surrey, and is also one of the most fecund of the beetles, laying up to 200 eggs at one time. It attacks the sapwood in softwoods, mostly in roofing timbers, but the holes are so large, between 6 and 10 mm diameter, that they often join up and severe structural weakening of timbers is likely. The exit holes are oval or ragged. The bore dust consists of sausage-shaped pellets which are gritty in feel. Identification can be confused with the jewel beetle, wood wasps, wasps or bees.

Any infestations should be reported to the local building control officer and to the BRE Timber and Protection Division.

Deathwatch beetle This attacks the sapwood and heartwood or partially decayed hardwoods, often oak and is associated with buildings of some age which use oak or elm. The timber needs to be to damp. The holes are 3 mm diameter and are randomly orientated, not necessarily in the direction of the grain. The bore dust is cream coloured and forms disc-shaped pellets which feel gritty between the fingers. From March or June they may be heard tapping out their mating call when they emerge from the exit holes.

12.4 Moisture content

The reduction of natural moisture content is critical to keeping timber in a stable condition, making it dry and unattractive to a range of species of fungi and insects that find it far more digestible when moist. Very broadly, the brown dry rots eat cellulose and the white rots digest both. The fungus cannot survive for longer than twelve months if dry conditions are established. As a general rule, timber is immune to attack if it maintains a moisture content (MC) below 20%. The average MC at time of erection varies according to the usage of the timber (see Table 12.7).

Table 12.7 Moisture content variation

Position of timber	Average moisture content with timber installed (%)	Minimum moisture content to be specified (%)
External use	18 or more	—
Covered but unheated	18	24
Covered but heated	16	21
Internal in continually heated building	14	19

Source: Table 1 BS 5268 Part 2: 1988

There is a long history in the use of timber preservatives to combat decay which extends back to the Middle Ages and includes washing timbers with lime solutions to kill organisms. Many of the compounds used currently in buildings have given cause for concern recently, and there are campaigns for withdrawing some of the more aggressive chemicals which are seen as a health risk, most obviously to operatives applying the treatment but also undesirable in places where people live and work. At present BS 5268 Part 5, *Preservative treatment for construction and timber*, published in 1977, still applies. Fungi and insects take in oxygen and breath out carbon dioxide, just like humans. Chemicals used to combat these living organism work by affecting the respiratory system, so in large quantities they will also affect people.

For advice on preservation and recommended treatments refer to Construction risks and remedies, timber decay part 2, *Architects' Journal* 15th October 1986.

Another effect of timber preservatives is in their potential to cause the electrochemical corrosion of metallic fasteners in timber, particularly those of steel. The compounds that have most effect are the salt-type preservatives which can initiate dissociation of metallic ions. In copper chrome arsenate (CCA) treated timber there is a general rule that no metal fixings should be applied within seven days of treatment. A similar effect can also be obtained with the use of fire retardants in timber which again may be salt compounds.

12.5 Corrosion and fixings

The main initiator for corrosion in fixings is the actual chemistry of timber. Water and acetic acid (from the hydrolysis of acetylated polysaccharides in wood) gives a general acidity which has a pH of 4.0, enough to cause corrosion mechanisms to become active. There must be sufficient moisture present for strong electrolyte activity and if the MC is about 20% and the relative humidity of the atmosphere about 75% these are conditions that will allow for the right volume movement of electrons. The timber surrounding metal fixings often becomes so damaged in this process that it causes *nail sickness*. This is more likely with wood freshly treated and with fixings inserted before the expiry of the recommended seven-day period. There is sufficient copper in solution in CCA to cause deposition on metals and initiate bimetallic corrosion.

'The corrosion loss of mild steel, bright zinc plated mild steel and aluminium alloy fasteners in damp wood have been increased by the copper chrome arsenate treatment.[2]

Zinc plated steel is at risk due to the thin coatings often applied. The test sample discussed in this paper had a measured plating thickness of only 12 microns.

Although hardwood is more acidic than sapwood, due to its density, the diffusion of copper and oxygen is slower than in sapwood and so corrosion is inhibited.

Aluminium fasteners are even more at risk: they were found to dissociate directly in a dilute solution (1%) of acetic acid.

$$Al = Al^3 + 3e^-$$

There are some basic points to watch. Timber should always be protected prior to fixing and should be stored above ground and under cover. Condensation is common in roof spaces and ventilation in roof spaces is essential. (See Building Regulation requirements.) Stainless steel fixings are not as much at risk due to the formation of chromic oxide on their surfaces (Cr_2O_3) although in further experiments[3] Simms and Button noted some minor indications of rusting, although the film surface under the microscope remained stable and did not contain microcracks which were seen in the mild steel, aluminium alloy and zinc samples. Generally stainless steel and copper alloys (except brass) seemed to be the least affected materials for fixing in CCA treated timber.

Sulphate ions were noted as being corrosion accelerators and may come either from the preservative or from 'atmospheric contaminants on the metal' or from diffusion through the wood of SO_2 (significant up to 2.3 mm below the surface.)

One important conclusion by Simms and Button was that chromate passivation treatments would help to control the phenomena of 'wet storage stain or white rust' on galvanized steel accessories, particularly plates and gussets. This was found while investigating chromate passivation in general. Samples were placed in *Cronak*[4] solution held at room temperature then washed, dried and aged for 24 hours before microscopic examination. The coating itself was defective but corrosion was only detected in the microcracks, indicating that more controlled and even applications would produce good protection.

12.6 Wood preservation

Preservatives and timber treatments work by affecting the life systems of living organisms, for example, the respiratory systems of beetles. It is logical that these compounds in large enough quantities will also affect people, and are regarded as toxic if applied in situations where there is inadequate ventilation. Some of the chemicals used are also accumulated in the body. Chemicals likely to be problematic include: pentachlorophenol (PCP) as a fungicide, banned in many countries and used externally in the USA only by qualified personnel; the anti-insecticidal treatments benzene hexachloride and dieldrin, banned in the USA since 1975 and to be banned from 1992 under an EC directive; tributylin oxide (TBTO) also an effective fungicide, banned for boat antifouling paint because of the effect on marine life and classified as a nerve poison; lindane a powerful insecticide (gamma-HCH or gamma-hexachlorocyclohexane) banned in Japan and the USA and other countries and recognized in the UK as endangering bats. These are all volatile organic compounds. There are now EC recommendations which outline alternative chemicals that should be used for this work and are less of a health hazard.

Toxic Treatments: Wood preservative hazards at work and in the home, published by the London Hazards Centre, gives a greater level of detail on the compounds used and their likely effects.

As a general rule, preservatives that present less risk are metal based inorganic compounds, e.g. copper napthanate, zinc napthanate and copper chrome arsenate. (Copper is fungicidal and arsenic is insecticidal.) Most preservatives contain *biocides*, a term that covers the inclusion of fungicides and insecticides. These compounds are still toxic and handling timber within two weeks of treatment is inadvisable, especially without gloves, as there can be absorption directly through the skin. In Australia, if CCA treated wood is to be used in children's playgrounds, it must stand for six weeks and then be hosed down before installation. For a detailed list see TRADA leaflet TBL 37: 1986, *Timber preservation*.

The least risk preservatives approved by the Nature Conservancy Council include permethrin and boron compounds.

The main strategy for buildings is that they should be primarily well ventilated spaces with an appropriate choice of timber, sufficiently durable for the situation, well seasoned with the correct moisture content. Preservatives

Table 12.8 Typical durability and treatability characteristics

Common name* or commercial designation†	Natural durability (heartwood only)‡	Treatability§	
		Heartwood	Sapwood
Balau (not including red balau)	D/VD	n/a	n/a
Beech, European	P	P	P
Cedar, western red (imported)	D	R	R
Douglas fir-larch (USA and Canada)	MD	R/ER	n/a
Elm, English and Dutch	ND	MR	P
Ekki	VD	ER	MR
Fir, Douglas (UK)	MD	R	MR
Greenheart	VD	ER	n/a
Hem-fir (USA and Canada)	ND	R	n/a
Iroko	VD	ER	n/a
Jarrah	VD	ER	n/a
Kapur, Malaysian	VD	ER	P
Karri	D	ER	n/a
Kempas	D	R	n/a
Keruing, Malaysian	MD	R	MR
Larch, Dunkeld (UK)	MD	R	MR
Larch, European (UK)	MD	R	MR
Larch, Japanese (UK)	MD	R	MR
Merbau	D	n/a	n/a
Oak, American white	D	ER	MR
Oak, European	D	ER	P
Opepe	VD	MR	n/a
Pine, Caribbean pitch	D	MR	P
Pine, Corsican	ND	MR	P
Pine, Parana	ND	MR	P
Pine, Scots	ND	MR	P
Radiata pine (New Zealand)	ND	R	P
Redwood, European	ND	MR	P
Southern pine (USA)	ND	MR	P
Spruce, European (UK)	ND	R	n/a
Spruce-pine-fir (SPF) (Canada)	ND	R	n/a
Spruce, Sitka (UK)	ND	R	MR
Sitka spruce (Canada)	ND	R	n/a
Teak	VD	ER	n/a
Western whitewoods (USA)	ND	R	n/a
Whitewood, European	ND	R	n/a

*The common names used in this table are defined in BS 881 and BS 589.
†Where a timber consignment is purchased under a commercial designation, e.g. hem-fir or spruce-pine-fir (SPF), the commercial designation should be referred to in this list, not the individual species within the consignment.
‡Durability grades are: VD — very durable, D — durable, MD — moderately durable, ND — non-durable, P — perishable. The durability of the sapwood of all species is either perishable or non-durable.
§Treatability groups are: ER — extremely resistant, R — resistant, MR — moderately resistant, P — permeable, n/a — data not available.

Source: Table 1 BS 5268 Part 5: 1989

should be used as a last resort. Unfortunately, the quality of timber used generally and cost factors in the choice of timber and availability mean that most timbers rely on some treatment. Lower quality timber which is poorly seasoned is used in building, and preservatives are used as a matter of course. Consequently timber treatments have become a growth industry in the last 40 years.

There are also genuine dilemmas in specifying timber. Despite the above strategy, the National House Building Council (NHBC) requires components to be treated (see the *NHBC Standards* 1991), particularly for structural timbers and external joinery. Building societies also make demands with regard to the protection of their properties. These policies by powerful national organizations could still conflict with section 2 of BS 5589:1978 *Use of preservatives not hazardous to people/pets*.

Table 12.9 Classification of risk categories

Risk category*	Risk of fungal decay	Need for preservative treatment
1	Where conditions of use involve negligible risk of fungal decay	Unnecessary
2	Where there is a low risk of fungal decay	Optional
3	Where experience has shown that there is a high risk of fungal decay	Desirable
4	Where timbers are exposed to a continually hazardous environment leading to an unacceptable risk of fungal decay	Essential
M	Where timbers may be exposed to attack by marine borers	Essential

*Where appropriate, these categories are used to define the risk of insect attack in tables 4 and 5 by substituting the words 'insect attack' for 'fungal decay'. These five risk categories are based on the European Harmonization Committee (EHC) classification (as published in the EHC Reference Document 1988 (p. 15)).

Source: Table 2 BS 5268 Part 5: 1989

Even after preservation treatment has taken place and months or years have passed, toxic vapours can still be released. In Germany, advice is given to seal timbers that have been treated with polyurethane resins or paint to prevent vapours being inhaled.

Tables 12.8, 12.9 and 12.10 give information relating to durability, treatability and risk categories depending on situation and use of timber in buildings.

If wood preservatives are to be used, note should be taken of the differing responses to treatment. For example, some timbers are very permeable and sapwood certainly more than heartwood due to the lower density of the material.

British Standard BS 5589: 1978 is the code of practice for the preservation of timber and is divided broadly into the sections given in Table 12.11.

The British Wood Preserving Association (BWPA) have their own code of practice for treatment called the *Timber specifications manual* and BRE Digests 201 *Wood preservatives, application methods* and 296 *Timbers: their natural durability and resistance to preservation treatment* are useful. For guidance on the natural durability of timber see *Timber preservation*, published by TRADA in 1986, from which Table 12.12 is an extract.

Degradation of timber by ultraviolet light and chemicals

The ultraviolet light component of sunlight has sufficient energy to destroy the chemical bonds in the lignin of timber. Deterioration shows as the timber loses its natural colour and turns grey and eventually becomes fibrous when only the celloulose structure is left unbonded. These loose fibres will eventually start to splinter. This is why protective coatings to wood have attained such importance in the building industry.

Chemical applications to timber

It may be necessary to coat timber with chemicals particularly to remove old coats of paint. These can cause damage to the cellulose structure, particularly if the timber is immersed in a bath.

Strong acids below a pH of 3 and alkalis above a pH of 9 will break down timber by destroying the lignin and leaving the cellulose giving a de-bonded fibrous appearance. This often happens as a result of excessive stripping in removing paintwork which will destroy the timber.

Fire protection

Most timber is classified as Class 3 in the surface spread of flame and this can be upgraded to Class 1 or even Class 0 (as laid out in BS 746 or the Building Regulations). Impregnation of timber with chemicals such as ammonium sulphate, ammonium phosphate, sodium borate, chlorides and bromides will inhibit the natural combustion of timber and there are a number of proprietory treatments. As these are often water soluble, care must be taken to ensure their protection and positioning. Any intumescent coatings must be kept intact, and it should be pointed out (in handbooks) to facilities or building managers that they are not to be replaced or painted over.

TRADA have published a short guide in their Wood Information sheet 4/11 introducing concepts of:

- Ignition and fire growth.
- The fully developed fire.

Recommended with this brief guide is a whole series of sheets which detail individual elements, from doors to walling systems, internal and external which have been tested to give *proven fire resistance*. The importance of these tests is that components are tested to take their designed structural loading after fire damage, and they are realistic, as elements are tested as they would be constructed.

The relevant parts of British Standard 476 are:

Table 12.10 Risk characteristics of timber used in buildings

Situation	Component	Risk category (for fungal decay unless otherwise stated)	Timber whose heartwood can be used without treatment	Minimum preservative treatments (√ — recommended, X — not recommended)								Remarks
				P and MR treatability group				R and ER treatability group				
				CCA	OS	Boron	Creosote (kg/m^3)	CCA	OS	Boron	Creosote (kg/m^3)	
1. Roof timbers referred to in the Building Regulations: Materials and Workmanship. Approved Document to support Regulation 7	Rafters, purlins, joists, wall plates, sarking, etc.	4C for house longhorn beetle	Hardwoods	P3 20 g/L	V/1 or M/10	√	X	P3 30 g/L	V/1 or M/10	√	X	See situations 2 and 3 for recommendations related to fungal decay risk
2. Pitched roofs (dry) (refer also to situation 1 if relevant)	(a) Rafters, purlins, joists, sarking, etc.	2C for common furniture beetle; 1C for fungal attack	The heartwood of all timbers except those whose heartwood is not well defined, e.g. spruce	P2 20 g/L	V/1 or M/3	√	X	P2 30 g/L	V/1 or M/10	√	X	
	(b) Tiling battens	2B	Moderately durable or better (MD, D or VD)	P8 15 g/L	V/1 or M/3	√	80 or M/3	P9 15 g/L	V/4 or V/3	√	80	(1) For immersion treatment, it is essential to loosen bands (2) Not recommended if roofing felt is present
	(c) Wall plates	2C	Non-durable or better (ND, MD, D or VD)	P2 20 g/L	V/1 or M/3	√	80	P2 30 g/L	V/4 or V/3	√	80	
3. Pitched roofs (risk of wetting) (refers also to situation 1 above if relevant)	(a) Rafters, purlins, joists, sarking, etc.	3C	Moderately durable or better (MD, D or VD)	P3 20 g/L	V/2	√	80	P7 30 g/L	V/4	√	80	
	(b) Tiling battens	2B	Moderately durable or better (MD, D or VD)	P8 15 g/L	V/1 or M/3	√	80 or M/3	P9 15 g/L	V/4 or V/3	√	80	(1) For immersion treatment, it is essential to loosen bands (2) Not recommended if roofing felt is present
	(c) Wall plates	3C	Moderately durable or better (MD, D or VD)	P3 20 g/L	V/2	√	80	P7 30 g/L	V/4 or V/3	√	80	
4. Flat roofs: cold (enclosed beams), valley gutters	Joists, firrings and timber deckboards	3C	Moderately durable or better (MD, D or VD)	P3 20 g/L	V/2	√	X	P7 30 g/L	V/4	√	X	In some situations, e.g. high internal humidity, this should be considered as 4C category (refer to situation 1 if relevant)
5. Flat roofs: warm, inverted (exposed beams)	Joists, firrings and timber deckboards	2C	Moderately durable or better (MD, D or VD)	P3 20 g/L	V/1	√	80	P7 30 g/L	V/4 or V/3	√	80	See remarks for situation 4

Source: Table 4 BS 5268 Part 5: 1989

Table 12.11 BS 5589 : 1978 Code of Practice for the preservation of timber

Section	Location of timber	Preservative type	Life (years)
2	External woodwork in buildings not in contact with the ground, i.e. external joinery, windows, cladding, doors, soffits and frames	—	30–60
3	Agricultural and horticultural buildings for equipment used in growing crops, storage and processing, depending on need	Use preservatives not harmful to plants and animals	20–50
4	In contact with sea water, piers, jetties and bridges	Use creosote/CCA which resists leaching	15 (sea water) 30 (fresh water)
5	Fencing timber, posts, etc. in contact with ground	Pressure creosote	40
	Telegraph poles, sleepers, etc.		20

Table 12.12 Natural durability of some timbers

Durability	Species	Ground contact life (years)
Perishable	Beech Birch	<5
Non-durable	Scots pine Sikta spruce Obeche	5–10
Moderately durable	Douglas fir Sapele	10–15
Durable	Western red cedar European oak	15–20
Very durable	Teak Greenheart	>25

Source: Timber preservation 1986 TRADA

Part 4 Non-combustibility for materials.
Part 5 Ignitability.
Part 6 Propagation test for materials.
Part 7 Surface spread of flame.

Non-combustibility of materials relates to the ability to withstand 750 °C in the initial stages of a fire. One advantage of solid timber is the predictable behaviour of charring over time.

Common softwoods over 420 kg/m^2 will char to about 40 mm in depth, below this density charring may extend to 50 mm. Hardwoods will generally not char to more than 30 mm.

12.7 Specification of timber

The specification of timber should give:

- Species.
- Section size (see BS 4471) for tolerances and length if critical.
- Stress grading for structural use to BS 4978 or Joinery grade to BS 1186.
- Moisture content (dependent on location).
- Method of preservation (if applicable).

Sources of supply and sustainable hardwoods

There is concern currently about the use of hardwoods focussed on the seemingly short-term policies of timber wholesalers which respond to market forces. Public opinion has been heightened by access to information revealing that policies on forest management differ enormously between countries. Ideally, when timber is specified, especially with regard to hardwoods, there should be consideration over the dwindling of rain forests, and the subsequent increase of carbon dioxide in the atmosphere as a contributer towards global warming. In addition, logging activities which remove substantial tracts of forest expose land to soil erosion. Land is destabilized, and major mud avalanches can lead to the silting of river systems and loss of life.

As a guide, it is better to specify timber by the country of origin as there is documentation held by the Timber Trade Federation (TTF)[5] and Friends of the Earth (FOE) which identify those countries that have a proven management policy. Forest management should enable a *farming* policy to take place allowing for an annual yield of timber which is controlled, leaving immature trees to grow for future harvesting. Sections of forest are identified and felling is not allowed to take place until boles reach a certain diameter, often when over 600 mm. As timber is a genuinely renewable resource, limitations on felling should be looked at in conjunction with good management.

However concerned the Western world may be about the specification of third world timbers, their consumption is not enough to put pressure on countries to follow renewable yield management policies. Most tropical hardwoods are

Figure 12.8 Bamboo fence showing junctions of three elements stabilized with one cross tie finished with a knot. The knot is a reminder of ties and obligations. Kiyomizu-dera Temple, Kyoto.

used in the country of origin and often only about 5% of their yield will be exported.

For example, in Malaysia 2.3×10^6 m^3 of wood is used in that country and only 1.25×10^3 m^3 exported to the UK. In Indonesia 9.2×10^6 m^3 is used internally and 1.0×10^3 m^3 exported to the UK. However, there are good examples of management and in Indonesia (Java) teak has been taken from a managed forest, initiated by the Dutch and now carried on by Indonesia.[6]

The TTF are trying to raise money to help with proper management but it would require awareness of these issues to be publicized in the countries of origin and perhaps worldwide condemnation of bad management practice followed by individual member state legislation before there are significant changes in attitude.

An example of using resources carefully is in Malaysia where rubber trees, originally an artificial plantation from South American seeds nurtured at Kew, are now being used for other purposes. Rubber trees after reaching between 25 and 28 years of age are over-mature for rubber production and were taken down and burnt to allow for new planting. The natural rubber industry has also declined with the rise in synthetic rubber production. So here is a tree which is an extremely dense hardwood with great potential. The timber is not durable and is liable to insect attack but extra research in Malaysia has improved delivery from forest to timber merchant and after being sawn, the timber is successfully used for furniture production.

12.8 Use of green timber and pole construction

There is a history and worldwide practice for using timber without processing it first and for using the forces of friction to stabilize joints by tying them. The direct use of timber without sawing is obviously energy saving, although still labour intensive. Traditional detailing of timber trusses in

Figure 12.9 The sorting of timber poles in Northern Honshu, Japan. There are a number of sections here that can be used directly in buildings without processing. Choice of the right size of the pole could pre-empt the careful shaping of timber into rectangular sections, stacking and seasoning to prevent warping, and then shaping them into ovals or rounds for rails, broom handles, etc. The orthogonal culture worldwide in the twentieth century is more limiting. Sawing can waste up to 40 per cent of a log.

Figure 12.10 Bamboo and ad hoc timber scaffolding in Bangkok, Thailand. Tying structural elements which use the very strong forces of friction are not common in Europe and yet these structures are used for constructing multi-storey buildings in the Far East. There is no diagonal bracing as the tying provides a sufficiently strong moment connection.

France and Japan also utilize whole timbers and by not cutting and exposing grain, the timber is more stable.

In the spirit of self-build construction, economizing on costs by using unseasoned or *green* timber is a possible option. Experimentation in the USA is well documented. (See *Low cost green lumber construction*, by Leigh Seddon, 1981, Garden Way Associates.) Wood bought from timber that has been freshly cut has a high moisture content, often up to 50%, but is often at least half the price of kiln-dried timber. The sections will be rough sawn and although it is usable, the designer must be aware and detail for the possible moisture changes that will take place. The timber should be air dried for as long as possible to at least 20% moisture content which may be between 1 and 6 months depending on the size. Timber needs to be stacked carefully on site and well supported to allow for air movement. If the support centres are too far apart the timber could deform through permanent cold bending by propping at too large spans.

Figure 12.11 Buildings using roundwood or pole construction in the School of Woodland Industry at Hooke Park, Dorset. Design team, Frei Otto, ABK and Buro Happold, 1985. (See *Innovation in the Woods*, *AJ*, 20th November 1985.) These buildings use 'thinnings' which would otherwise be sold for pulp. This roof is a catenary structure with ridge cable; the rafters lie from the ridge to the eaves perimeter beam and the natural bending gives an attractive roof. Special jointing techniques were developed with Bath University.

General points to make include ensuring that subframing around windows and glazed doors are made of seasoned timbers otherwise cracks will occur. Floor boards can be laid loose until seasoned and then fixed. To ensure that movement is minimized section sizes are larger than normal.

Round pole construction

A major example of pole construction undertaken by the Parnham Trust was the Workshop Building, again with the team of Frei Otto and Burro Happold in conjunction with the University of Bath.

Using the thinnings from the local wood the Parnham Trust decided to develop a building technique which made use of essentially waste material left over from the selective thinning of the managed woods and demonstrated 'learning to use the resources you have' (John Makepiece). This lumber was normally sold off by weight as pulp for paper and other products. The wood yields a high grade stock of Douglas fir and Norwegian spruce. The thinnings are between 30 and 35 years old. Their size is approximately 6–8 m in length with a bole section that varies from 150 mm at the base to 60 mm at the tip. It was decided to use these whole sections to form a large enclosure by bending them when they were still green, which is 30% moisture content, the theoretical fibre saturation point. The bending radius of the timbers is dependent on the diameter of the timbers. The poles were not really slender enough for the curvature required, the mid-diameter was felt to be the right size. A mechanism was set up by the engineers which pulled the poles over and strain guages were used to monitor the degree of deformation related to the permissible stress calculated. In the end this procedure proved to take such a long time that a decision was taken on site (largely by the foresters) to bend the poles virtually by eye and hand, and to reject those that showed signs of fracture or unacceptable deformation. This allowed for the *pulling* of sections to take place within a much shorter timespan, i.e. two days. Timbers were wrapped in polythene to slow the drying out process. As green timbers are bent hydraulic pressure from fluid in the sap walls can cause rupture of the sections.

Gradual bending can give a pre-stressing to the timbers, which then can be progressively bent even further. Setting of the required shape is a combination of using restraint to sections by anchoring them to restrict their elastic behaviour, which would cause them to spring back to their original profile once the stress was released, and careful drying out. After setting the thinnings should be kept impervious to any moisture.

In bending, approximations can be made with regard to the likely radius to be achieved by using the formula:

$$R = 50S$$

where R = radius of curvature, S = thickness of section.

An example of cold bending using this formula shows that a 25 mm thick timber will bend to a radius of

Figure 12.12 The main workshop building at Hooke Park showing the use of poles held in a concrete plinth bent over and jointed by lapping and binding with glass fibre tape and epoxy resin.

approximately 1.2 m.

As the section size of thinnings is variable, this means the radius will vary and give either the limitations of a radius prescribed by the larger section, or a parabolic shape which would optimize the bending of the whole section. As designers are unused to dealing with tapered sections, common assumptions in detailing overlook the nature of the diminishing section.

These particular timbers were treated in a preservative solution by immersion in a boron bath at 55 °C. Thinnings are immersed in bundles and the treatment is then repeated after 72 hours. The thinnings are then air dried under polythene for four weeks before being subjected to bending. There is some crystallization from the boron salts on the surface of the timber.

Pole thinnings and whole-tree utilization

If the utilization of timber is carefully thought through this can affect the design of components. One company in Sweden has been successfuly making timber beams and trusses which use ex 500 × 50 mm sections rebated into which is glued a vertical panel. Four of these sections can be cut from a timber thinning with a minimum bole diameter of 150 mm. All of these sections will be quarter sawn and more stable. The waste is then processed into fibres, mixed with a resin and compressed to make the vertical infill member. The full-scale utilization of timber thinnings in this way is an example of how timber design and the processing of the material can be optimized to yield a very efficient product.

This is also an example of *whole tree utilization* built on the new technologies of using wood waste.

References

1 J D Barrett 1981 Fracture mechanics and the design of wood structure. In *Philosophical Transactions* Royal Society, London: pp. 217–26.
2 D W Simms and H E Button 1985 Corrosion of certain metals in CCA treated timber. *Corrosion Prevention and Control* August.
3 D W Simms and H E Button 1985 Electrochemical measurement on metals in CCA treated wood. *Corrosion Prevention and Control* August.
4 'Cronak' solution consists of 200 g $Na_2Cr_2O_7.2H_2O$ and 9 ml H_2SO_4 (s.g. 1.84 in 1 litre of water).
5 Timber Trade Federation, Claverville House, 26/27 Oxendon Street, London SW1 4EL; Tel. 0171 839 1891.
6 Simon Counsell 1990 *The good wood guide* Friends of the Earth publication, London.

13 Artificial composites: Timber-based boards and slabs

The use of board materials has steadily increased in the building industry as they provide a wide range of components which can be used for structural and non-structural applications. They speed up the building process as the fixing of slabs replaces the laborious work of fitting small elements on site. BRE Digest 323 provides a guide to *Selecting wood based panel products* (1987). There are a series of charts which lists all the different boards available according to their location in a building and the likely performance needed.

All timber-based boards and slabs are composites because of their combination with various adhesives. They may be laminar composites, with sections or sheets of timber that are glued together in line with great directional strength, or particle composites which use waste chips, or fibres or flakes of material. They are normally characterized by their bonding and composite, which will be chosen according to their degree of exposure.

BS 6100 Section 4.1:1984 is a guide to forest products with a full glossary to all terms used. TRADA Wood Information sheet WI 23, 1984, is useful as an *Introduction to wood-based panel products*.

Plywood and laminated timbers are regarded as structural timbers and come under BS 5268. Chipboards and fibreboards do not as they lack continuity of structure; it is fairly easy for fracture or failure to occur around particle boundaries.

Figure 13.1 Plywood used directly as external cladding material for a timber/stone workshop in Hooke Park, Dorset.

13.1 Laminated composites

Plywood

Plywood is made from veneers from logs which have been prepared by soaking in hot or cold water and are then rotated against sharp cutting blades which peel the log into prescribed thicknesses. The sheets produced are then dried, spread with adhesive and then bonded with the grain of each veneer assembled at right angles to the previous layer. This is carried symmetrically to a core veneer, so plywoods will always be made up of an odd number of plys and British hardwood plywood would give the following configuration:

7 mm	3 ply
12 mm	7 ply
19 mm	9 ply
25 mm	11 ply

Thicknesses may be slightly less for variations of veneer thickness.

Plys are counterbalanced around the core sheet. They are pressed and heated to facilitate bonding. If preservatives are needed, they might interfere with bonding and also the application of exterior finishes, so compatibility assurances should be sought from manufacturers. Their moisture content will be between 6 and 14%. Each ply in coniferous plywoods should not exceed 6.5 mm, and in other plywoods the inner ply should not exceed 5 mm and the outer ply should not exceed 3.5 mm. The outer veneers should not be less than 0.6mm.

Relevant British Standards are:

BS 1088 and 4079 *Specifications for plywood for marine craft*, 1966 and 1981.

BS 1203:1979 *Specification for synthetic resin adhesives*. The adhesives used in this standard can be:

- Synthetic resin adhesive which may be phenolic or aminoplastic.
- Phenolic resin from the condensation of a phenol (phenol, cresol, xylenol, resorcinol) with an aldehyde to produce formaldehydes or furaldehyde.
- Aminoplastic resins, commonly urea or melamine combined with formaldeyde.

Hardeners may be used as curing agents. These still retain types characterized by letters. For example:

WBP Weatherproof and boil-proof Resistant to weather generally, micro-organisms, cold and boiling water, steam and dry heat. (Phenol formaldehyde bonded.)

BR Boil resistant Although reasonably resistant to weather and to boiling water, they are not as durable as plywoods made with WBP adhesives. They are still resistant to attack by micro-organisms. They are ideal for indoor use in wet conditions. (Melamine urea formaldehyde.)

MR Moisture resistant and moderately weather resistant They deteriorate externally after a few years and cannot resist boiling water. Resistant to attack by micro-organisms. (Urea formaldehyde bonded — Strong blend.)

INT Interior These plywoods have adhesives that are resistant to cold water but may suffer attack from micro-organisms. (Urea formaldehyde — Weak blend.)

Refer to BS 6566 Plywood Parts 1–8 for general information.

BS 5268 Part 2:1988 *Structural Use of Timber*. This standard outlines plywoods that are all bonded with exterior type adhesives, but an additional note is given on durability with regard to the choice of hardwood veneers for exterior use which should also be treated with preservative. Complete details are given of:

- American construction and industrial plywoods.
- British hardwood plywoods.
- Canadian Douglas fir and softwood plywoods.
- Finnish birch, birch-faced and conifer plywoods.
- Swedish softwood plywood.

The face grain of ply should be laid parallel to the span, otherwise if it is laid perpendicular the effective strength will be reduced to about one third. For example, exact ratios see Table 13.1 (and refer to tables 39–52 in BS 5268 Part 2 1988). Do not confuse this condition with decorative veneered-faced plywood which may be only 1 mm thick and lie at right angles to the structural ply sheathing below.

Blockboard and laminboards

These are made by aligning strips of timber not over 25 mm in width that have been cut from planks which are often the residue cores from plywood manufacture. Laminboards use strips that are up to 7 mm in width and are continuously glued. They are generally regarded as a superior product and are a better base for high quality surface finishes or veneers. The moisture content is higher than for plywoods and approximates to $10 \pm 2\%$. These strips are not necessarily continuous throughout the length of the board. They are then joined together with the top and bottom faced with a glued veneer which runs at right angles to the direction of the strips. If they are faced with one veneer on either side they are of three-ply construction, with the veneers not less than 2.5 mm or more than 3.6 mm in thickness. An additional veneer can be added at right angles to give five-ply construction, which should not be less than 2.5 mm or greater than 5 mm. BS 3444:1972 is used by manufacturers as a quality control standard. The timbers and veneers are glued together with urea formaldehyde resins which are not as durable as the WBP and BR grades of adhesive for plywood. Consequently blockboards should not be used in exposed situations and are better for internal use, although the specification can be changed if the durabilty requirements are made known to the manufacturers. Bonding grades to BR can be supplied as well as MR and INT.

Although veneers are graded they are often a source of consternation on site. If the quality of the veneer is critical, and knots are unacceptable, it should be selected. Dependent on the exposure of the plywood it should be rated in terms of durability, and this will affect the choice of veneers. Preservatives will have to be used to improve the durability of some veneers and BS 6566 Part 7:1985 gives an outline of the grades of durability and preservative treatments involved.

Veneering terms As most of these slab materials can be veneered there are four basic methods for cutting veneers which can be specified.

Sliced quarter cut This is a vertical radial cut through the log with sawn sheets giving a parallel figure showing the vertical direction of the grain. There is wastage because of the relatively small end pieces. It is a stable

Table 13.1 Dry grade stresses and moduli for American construction and industrial plywood

Type and direction of stress and modulus	Stress or modulus for a given nominal thickness with number of piles in parentheses (N/mm^2)										
	9.0 mm (3)	9.5 mm (3)	12.0 mm (4)	12.0 mm (5)	12.5 mm (4)	12.5 mm (5)	15.0 mm (4)	15.0 mm (5)	18.0 mm (5)	19.0 mm (5)	19.0 mm (6)
*Extreme fibre in bending**											
Face grain parallel to span	9.64	9.21	9.55	9.55	7.70	9.25	8.09	7.82	7.22	6.95	6.98
Face grain perpendicular to span	4.34	7.37	5.45	5.08	9.62	6.96	4.37	7.90	5.59	8.99	8.62
*Tension**											
Parallel to face grain	3.41	3.15	3.36	4.72	2.38	4.89	3.80	3.97	4.36	4.08	4.18
Perpendicular to face grain	3.96	5.62	4.68	3.67	6.52	4.02	3.35	4.14	4.17	5.01	4.33
*Compression**											
Parallel to face grain	4.96	4.60	4.88	5.83	3.49	6.08	4.69	4.94	5.39	5.06	5.19
Perpendicular to face grain	4.90	6.99	5.79	4.54	8.13	5.00	4.14	5.14	5.25	6.24	5.38
Bearing											
On face	2.67	2.67	2.67	2.67	2.67	2.67	2.67	2.67	2.67	2.67	2.67
Rolling shear in plane of plies											
In face veneer / In back veneer / At first glue line	0.45	0.45	0.45	0.45	0.45	0.45	0.45	0.45	0.45	0.45	0.45
Transverse shear											
Bending: face grain parallel to span	0.50	0.55	0.52	0.55	0.58	0.59	0.53	0.60	0.55	0.62	0.62
face grain perpendicular to span	0.83	1.05	0.92	0.30	1.20	0.36	0.28	0.40	0.30	0.40	0.41
Panel shear											
Parallel and perpendicular to face grain	1.37	1.12	1.54	1.81	1.29	1.67	1.43	1.33	1.17	1.06	1.30
Modulus of elasticity in bending											
Face grain parallel to span	9096	8700	9013	8502	7250	8200	7195	6950	6427	6200	6200
Face grain perpendicular to span	1377	3400	2058	2625	5150	4400	2049	5400	3044	6500	6200
Modulus of elasticity in tension and compression											
Parallel to face grain	6031	5550	5928	7084	4200	7350	5701	5950	6550	6100	6250
Perpendicular to face grain	5951	8450	7033	5510	9800	6050	5034	6200	6261	7500	6500
Shear modulus (for panel shear)											
Parallel and perpendicular to face grain	630	630	630	630	630	630	630	630	630	630	630

*For panels less than 600 mm wide, the bending, tension and compression stresses should be reduced in proportion to their width, commencing with no reduction at 600 mm to 50% at 200 mm and less.

Source: Table 41 BS 5268 Part 2 : 1988

quarter sawn cut. Veneers are easily matched to give a homogeneous appearance.

Sliced crown cut These veneers are produced in the same way that *through and through* sawing produces planks. The veneers are not as stable and can *cup*. Veneers have to be carefully matched and will show alternating banding, placed symmetrically around the axis of the veneer face.

Rotary cut Here the log is peeled by rotation against a fixed blade. This method is used for producing the inner veneers of structural plywood, there are no joints in the sheet face and the figure becomes irregular, reflecting the differing growth rings being cut through.

Eccentric round cut This uses the same process as the rotary cut but only on one quarter of the log. If the four quarters are cut and bundled matching can more systematic. This is ideal for burr walnut.

When choosing veneers samples should be looked at that have been polished or sealed. The colour rendering will inevitably change as the surface film of any finish is sufficient to absorb and reflect light, and surfaces may have to be artificially lightened before an applied clear finish deepens the colour.

13.2 Fibre composites

Fibre composite boards

These include a wide range of boards from the low density insulating boards to medium density fibreboards and to dense hardboards. Most boards are imported. The Fibre Building Board Organization Ltd can provide advice (FIDOR 1, Hanworth Road, Feltham, Middlesex, TW13 5AF, Tel. 0181 751 6107) and BS 1142 covers these materials in three parts:

Part 1 *Methods of test.*
Part 2 *Medium board and hardboard.*
Part 3 *Insulating board or softboards.*

All fibreboards are made from solid wood which has been broken down into fibres that are reconstituted under pressure and heating. Waste material is chipped, softened in steam and then ground. Water is then added to form a loose mixture and this is laid on to a moving wire mesh which is subjected to suction to remove water and rolling to compact and *felt* the fibres. Hot pressing is used to compress the boards to medium or higher densities.

Hardboards type S These are dense boards at over 800 kg/m^3, smooth-faced with a mesh pattern on one side. The smooth face has a good base for laminated or coated finishes.

Tempered hardboards are even more dense at over 960 kg/m^3. There are two grades, TE and TN. TE is the superior product with regard to strength. Both are produced in thicknesses ranging from 3.2 to 12.7 mm.

Mediumboards Mediumboards are available in two grades: type LM (low density) 350–560 kg/m^3 which can be used for pinboards and linings ranging from 6.4 to 12.7 mm in thickness; type HM (high density) 560–800 kg/m^3 which are used for wall linings and partitions and range from 8 to 12 mm in thickness.

Softboards Soft or insulating board has the lowest density, under 240 kg/m^3, with thicknesses varying from 12 mm. It is used for thermal insulation and as pinboard. It can be provided impregnated with bitumen for roof sarking or as a floor underlay.

Medium density fibreboard This board differs from the other fibreboards by the addition of resins, usually urea formaldehyde and wax emulsion, the same bonding agents used for chipboard, and cured by the application of heat and pressure but on steel plates. This gives a high surface quality finish to the board with greater potential for fine work which can be used for joinery.

Flakeboards, waferboards and orientated strand boards

Logs are processed by reduction to wafer flakes which are dried, blended with phenol formaldehyde or melamine urea formaldeyde and distributed onto a mat-forming process which orientates the flakes, giving them either a random orientation or directional orientation and consequently greater strength. They are pressed and cured and are used in situations similar to plywood. They are not covered by a British Standard. They originated in North America as a product and are made from either hardwoods (aspen) or softwoods (pine).

Compressed straw building slabs

These slabs are not used so much today because of the advancing technologies of other board materials. BS 4046 Part 2:1971 is still relevant. They were popular as building boards for roofing and cladding because of their good insulative value with a thermal conductivity of 0.108 W/m K relative to other materials available at the time, now superseded by other more efficient insulation materials which can be used for light traffic on roofs. They are made from natural straw compressed with heat and pressure with smooth linings without any bonding agent. The co-ordinating slab sizes are in widths of 1200 mm with a 50 mm thickness and in lengths of 1800 mm, 2400 mm, 2700 mm, 3000 mm and 3600 mm.

They must be well protected from damp and placed in situations where they are kept dry and well ventilated. It is not unknown for slabs in the right conditions to sprout and provide an unexpected garden amenity at roof level. Flat roofs should be supported at 600 mm centres.

Wood-wool

Wood-wool slabs are made from long wood fibres of seasoned, chemically impregnated timber which are coated with Portland cement and then bonded under pressure. Long span flooring is possible by using channel edge reinforcement to the panels. These slabs can be used for roof decking, floors, permanent shuttering (see BS 3809:1971 *Wood-wool permanent formwork and infill units*), partitions and exterior cladding and have been proved to be a durable material. They can be provided with a slurry of cement for a good surface finish. (See BS 1105 and BS 3809:1971.) There is an advantage in using wood-wool for shuttering because of the thermal insulation it provides which will help in the curing process.

These slabs are made in standard widths of 600 mm, the effective spanning width for a depth of 50 mm and the dimension at which supporting structural joists should be fixed. Additional thicknesses of 75 and 100 mm are also available. The 100 mm thickness will span 1200 mm. Much longer spans can be achieved by the use of integral metal channel stiffeners at the sides of the slabs which increase the maximum span to 4000 mm.

Wood-wool achieves Class O for fire propagation under Part 6 of BS 476:1968 and Class 1 for surface spread of flame under BS 476 Part 7:1971. Pre-screeded wood-wool has achieved one hour fire resistance in accordance with BS 476 Part 1:1953. (See BS 1105:1981 *Wood-wool cement slabs for three grades of panel rated according to mechanical performance*.)

13.3 Particle composites

Chipboard

Chipboard has been developed since the second world war as a repercussion of the development of synthetic adhesives. It uses waste timber which includes residues from working timber as well as thinnings from forests. It is mechanically reprocessed to a controlled chip size. Their moisture content is then reduced to 3% by tumble drying. They are blended normally with urea formaldehyde resins which may be graded. Chip sizes may be selectively dropped onto a mat. Often finer paricles are on the outside of the sheet material with a coarser body fill in the centre to obtain a good surface quality finish which is smooth to receive bonded melamine sheets. The sheets are compressed and heat cured to set the adhesive. BS 5669:1979 *Specification for wood chipboard and methods of test for particle board* is applicable. There are four main types:

Standard Type 1 to BS 5669 These have a good surface quality for the application of veneers or coatings.

Flooring Type II to BS 5669 These are higher-strength boards for flooring often machined with tongued and grooved edge details for easy laying.

Moisture resistant Type III to BS 5669 These boards have melamine fortified urea formaldehyde adhesives or phenol formaldehyde with a density of 680–750 kg/m^3. They perform better in damp situations, although they will not withstand prolonged exposure.

Moisture resistant/flooring Type II/III to BS 5669 These boards combine the properties of board Types II and III.

Chipboards are known to suffer considerable loss of strength if they absorb water. After exposure to a relative humidity of 87% at 27 °C for 25 weeks and then reconditioned in an atmosphere of 65 °C, they were found to attain only 66% of their initial strength. The thickness of the board increased by 9%.

Flooring grade chipboard should always be used for that specific situation and not exposed to water spillage which may occur in kitchens or bathrooms. If chipboard is to be used then it must be fully tanked. Vinyl tiles are not thought adequate protection. Apart from specifying the correct grade of chipboard and bonding agents, pre-waxed sheets do help in providing some protection against moisture.

Wood–cement particle board

Wood–cement board contains only between 20 and 30% wood which is processed into fine particles directly from logs. These particles are mixed with water, cement and other chemicals, spread onto caul plates, clamped during initial curing, cured in controlled conditions in setting chambers and then stored in a warehouse for 18 days. They are often three-layerboards with particles in the centre and long fibres (30 mm) on the outer faces. The boards are finished with cement-rich outer layers. They are very heavy with a density varying between 1000 and 1200 kg/m^3. They have great dimensional stability, have excellent fire-resistant properties and will weather well. There is no British Standard for this product. They are best described as *render slabs* and are good exterior cladding materials. The timber content strengthens them and relieves the brittleness associated with pure cement slabs. They would provide excellent sound insulation in the refurbishment of flats.

Manufacturers to contact include Pyrok Building Products

Figure 13.2 Timber planking in Rye, Sussex, showing cold bending of timber onto a stud frame behind giving a corner detail to a warehouse building. The technology for this detail has a clear reference back to the local naval architecture. It also avoids the making of a more difficult corner junction.

Figure 13.3 Timber bending in a Thonet chair in the Museum d'Orsey collection, Paris.

Ltd and Torvale Building Products Ltd. The boards are marketed as an alternative to asbestos cement boards with thicknesses that vary from 6 to 40 mm. They can be sawn and drilled with tool wear that has been proved to be less than for chipboards. They can also be clad with decorative laminates for finishes as well taking paint coatings directly, but compensating finishes must be applied to both sides. They can also be wallpapered. Screw fixing is normal practice, although they can also be bonded to brick, wood or metal (if a primer is used) using resorcinol formaldeyde or in protected environments to wood with a catalyzed poly(vinyl acetate). They comply with BS 476 for Parts 5 and 6 (Class O) and have attained Class 1 in Part 7. Their K value is 0.23 W/m K. If cover strips of the same material are used on 75 mm timber studs before overfixing of these boards, two hours fire resistance can be achieved, and three hours with 150 mm deep steel stud partitions. Tongued and grooved slabs are made for flooring which if floated on 25 mm thick mineral fibre quilting will give good sound insulation.

BRE Information paper 4/83 *Wood cement particleboard — a technical assessment*, by J M Dinwoodie and B H Paxton, gives an extensive history of the material and indicators for its long-term performance.

Hollow core slab boards

It is worth mentioning that *slab technologies* have been developed by door manufacturers who have devised ways of producing hollow core constructions that are strong with infills that are core strips of cardboard in honeycomb, chevron or spiral configurations. These give rigidity to the board because of their depth but are lightweight because of their economic strength:weight ratio. This is one of the few advances in building that have been imitated in aircraft engineering where honeycomb systems use aluminum expanded cells sandwiched between aluminium alloy sheets. Their strength relies on good bonding between the edge of the infill material and the enclosing slab. (See *MBS*: *Internal components for ranges of doors*, and BS 4965:1983

Figure 13.4 Extreme bending showing distortion beyond a smooth curvature, possibly in growth. Phone support Japan. On old road to Edo between Nakatsugawa and Magome, Northern Nagono-ken.

Figure 13.5 The making of a fishing boat in Spetse, Greece: (top) the main ribs shapes need timber to be cut away at critical points to assist curvature; (bottom) the wastage in cutting the curved boat form. This is a technique also used in the making of coffins.

Decorative laminated plastics sheet veneered boards and panels.)

13.4 The bending of timber

Bending timber is quite common for all sorts of applications and ranges from domestic artefacts such as baskets, chairs and sports equipment to boats and building. There are two methods of bending, either when the timber is cold or when heated to form tighter radii of curvature. The degree of bending depends on the section size of timber, the size of timber and the moisture content. Timber can be bent directly under the application of heat but this causes discoloration which can be seen in most cane furniture. Generally the application of moisture and heat in the combination of steam

Figure 13.6 Side of storage house in Hida Takayama village. This shows the Japanese preference for holding the timber boards in position with other horizontal timbers, keeping them in position without nails. The butt jointed details are kept for ventilation.

puts timber in the best state for bending. Heating to boiling point is normally adequate.

The application of steam prevents discoloration and is a common method for bending especially for repetitive shapes which can then be bent around formers in a vacuum chamber. Failures in bending timber are often due not to over-stressing which gives a straightforward compression failure and buckling of the cell walls, but from the hydraulic pressure of water in the cells which can then rupture the cell walls, particularly in green timber.

Cold bending does not give a permanent set as the timber is still quite elastic, and so it has to fixed back permanently to a rigid frame, for example, in boat building where planks are secured back to the ribs. In making a curve in timber it can also increase its structural efficiency as stiffness is produced by the effective depth of the whole shape. This is a property not over-exploited by architects but can lead to quite elegant walls and doors.

Figure 13.7 Knots in timber are inevitable due to the nature of the natural material. These planks have cover strips of battens which give a strong rhythm to the wall. Kyoto, Japan.

A rough formula for cold bending is:

$$\frac{\text{Thickness}}{\text{Radius of curvature}} = 0.02$$

This is not exact but it does show that a 25 mm thick plank cannot be bent to a radius much less than 1200 mm. Laminated bending allows for the building up of quite large sections with tight radii. These are common techniques for large scale timber sections for portal frames where the introduction of a controlled curve is a stress relieving geometry.

For greater detail see *The wood bending handbook*, by W C Stevens and N Turner, HMSO, 1970.

There are some differences in the cold bending of laminated timber sections as against pole sections. To start with the timber can be selected to eliminate most structural weakening features, e.g. knots, wane and fissures generally. Whole structural members can be profiled to allow for changes of section where there would be greater bending (for example, in the design of portal frames, allowing for a thickening of the shoulders.)

13.5 Natural timber cladding

It is worth re-looking at timber as a natural cladding material as the variations for its use internally and externally are great.

Boarding should be not less than 16 mm in thickness apart from feather edges which should be a minimum of 6 mm. External ply cladding should be minimum of 9 mm in thickness although Scottish regulations require a minimum of 21 mm. (See *External timber cladding* TRADA WI sheet 1-20 1987.)

Figure 13.8 This timber house in Northern Nagano-ken is deteriorating but shows the variation in a window wall system that gives choice, with a sliding cover screen with a repetitive vertical patterning that is known to induce calm. The paper windows (when whole) give a diffuse quality of light to the interior.

Index

This index is presented in three sections: *Architects*, *Buildings* and *General*. Page numbers shown in **bold** type in the *Architects* and *Buildings* sections give references to illustrations.

Architects

Buildings

General